Geohydrology, Simulation of Regional Groundwater Flow, and Assessment of Water-Management Strategies, Twentynine Palms Area, California

By Zhen Li and Peter Martin

Prepared in cooperation with the U. S. Marine Corps, Department of Defense

Scientific Investigations Report 2010–5249

U.S. Department of the Interior
U.S. Geological Survey

U.S. Department of the Interior
KEN SALAZAR, Secretary

U.S. Geological Survey
Marcia K. McNutt, , Director

U.S. Geological Survey, Reston, Virginia: 2011

For more information on the USGS—the Federal source for science about the Earth, its natural and living resources, natural hazards, and the environment, visit http://www.usgs.gov or call 1-888-ASK-USGS

For an overview of USGS information products, including maps, imagery, and publications, visit http://www.usgs.gov/pubprod

To order this and other USGS information products, visit http://store.usgs.gov

Suggested citation:

Li, Zhen, and Martin, Peter, 2011, Geohydrology, simulation of regional groundwater flow, and assessment of water-management strategies, Twentynine Palms area, California: U.S. Geological Survey Scientific Investigations Report 2010–5249, 106 p.

Contents

Figures

Tables

Conversion Factors, Vertical Datum, Water-Quality Information, Abbreviations, and Well Numbering System

Conversion Factors

Multiply	By	To obtain
inch (in.)	2.54	meter (m)
foot (ft)	0.3048	meter (m)
mile (mi)	1.609	kilometer (km)
acre	0.4047	hectometer
square foot (ft^2)	0.09290	square meter (m^2)
square mile (mi^2)	2.590	square kilometer (km^2)
acre-foot (acre-ft)	0.001233	cubic hectometer (hm^3)
acre-foot per year (acre-ft/yr)	0.001233	cubic hectometer per year (hm^3/yr)
cubic foot per second (ft^3/s)	0.02832	cubic meter per second (m^3/s)
gallon (gal)	0.003785	cubic meter (m^3)
million gallons (Mgal)	3,785	cubic meter (m^3)
gallon per minute (gal/min)	0.06309	liter per second (L/s)
gallon per minute per foot [(gal/min)/ft)]	0.2070	liter per second per meter [(L/s)/m]

Concentrations of chemical constituents in water are given either in milligrams per liter (mg/L) or micrograms per liter (µg/L).

Vertical Datum

Vertical coordinate information is referenced to the North American Vertical Datum of 1988 (NAVD 88).

Horizontal coordinate information is referenced to the North American Datum of 1983 (NAD 83).

Altitude, as used in this report, refers to distance above the vertical datum.

Abbreviations

m	meter
MCAGCC	Marine Corps Air Ground Combat Center
MCL	Maximum Contaminant Level
NWIS	National Water Inventory System
MF2K	MODFLOW-2000
RMSE	Root mean square error
USGS	U.S. Geological Survey

Well-Numbering System

Wells are identified and numbered according to their location in the rectangular system for the subdivision of public lands. The identification consists of the township number, north or south; the range number, east or west; and the section number. Each section is further divided into sixteen 40-acre tracts lettered consecutively (except I and O), beginning with 'A' in the northeast corner of the section and progressing in a sinusoidal manner to 'R' in the southwest corner. Within the 40-acre tracts, wells are sequentially numbered in the order they are inventoried. The final letter refers to the base line and meridian. In California, there are three base lines and meridians: Humboldt (H), Mount Diablo (M), and San Bernardino (S). All wells in the study area are referenced to the Mount Diablo base line and meridian (M). Well numbers consist of 15 characters and follow the format 02S007E02C001S. In this report, well numbers are abbreviated and written 2N/7E-02C1. Wells in the same township and range are referred to by only their section designation, 2C1. The following diagram shows how the number for well 2N/7E-02C1 is derived.

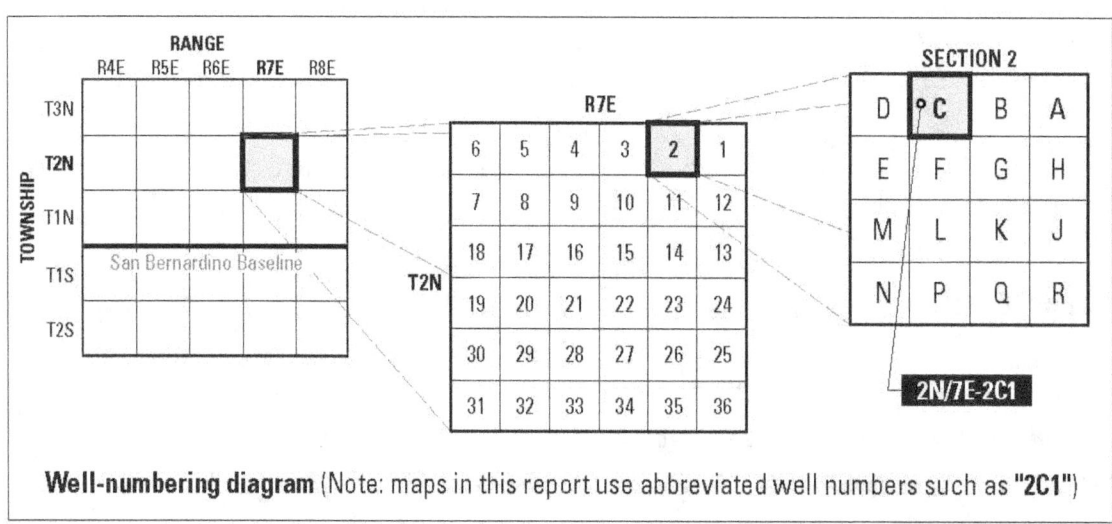

Well-numbering diagram (Note: maps in this report use abbreviated well numbers such as **"2C1"**)

Geohydrology, Simulation of Regional Groundwater Flow, and Assessment of Water-Management Strategies, Twentynine Palms Area, California

By Zhen Li and Peter Martin

Abstract

The Marine Corps Air Ground Combat Center (MCAGCC) Twentynine Palms, California, overlies the Surprise Spring, Deadman, Mesquite, and Mainside subbasins of the Morongo groundwater basin in the southern Mojave Desert. Historically, the MCAGCC has relied on groundwater pumped from the Surprise Spring subbasin to provide all of its potable water supply. Groundwater pumpage in the Surprise Spring subbasin has caused groundwater levels in the subbasin to decline by as much as 190 feet (ft) from 1953 through 2007. Groundwater from the other subbasins contains relatively high concentrations of fluoride, arsenic, and (or) dissolved solids, making it unsuitable for potable uses without treatment. The potable groundwater supply in Surprise Spring subbasin is diminishing because of pumping-induced overdraft and because of more restrictive Federal drinking-water standards on arsenic concentrations. The U.S. Geological Survey, in cooperation with the MCAGCC, completed this study to better understand groundwater resources in the area and to help establish a long-term strategy for regional water-resource development.

The Surprise Spring, Deadman, Mesquite, and Mainside subbasins are filled with sedimentary deposits of Tertiary age, alluvial fan deposits of Quaternary-Tertiary age, and younger alluvial and playa deposits of Quaternary age. Combined, this sedimentary sequence reaches a maximum thickness of more than 16,000 ft in the Deadman and Mesquite subbasins. The sedimentary deposits of Tertiary age yield a small amount of water to wells, and this water commonly contains high concentrations of fluoride, arsenic, and dissolved solids. The alluvial fan deposits form the principal water-bearing unit in the study area and have a combined thickness of 250 to more than 1,000 ft. The younger alluvial and playa deposits are unsaturated throughout most of the study area. Lithologic and downhole geophysical logs were used to divide the Quaternary/Tertiary alluvial fan deposits into two aquifers (referred to as the upper and the middle aquifers) and the Tertiary sedimentary deposits into a single aquifer (referred to as the lower aquifer). In general, wells perforated in the upper aquifer yield more water than wells perforated in the middle and lower aquifers. The study area is dominated by extensive faulting and moderate to intense folding that has displaced or deformed the pre-Tertiary basement complex as well as the overlying Tertiary and Quaternary deposits. Many of these faults act as barriers to the lateral movement of groundwater flow and form many of the boundaries of the groundwater subbasins.

The principal recharge to the study area is groundwater underflow across the western and southern boundaries that originates as runoff in the surrounding mountains. Groundwater discharges naturally from the study area as spring flow, as groundwater underflow to downstream basins, and as water vapor to the atmosphere by transpiration of phreatophytes and direct evaporation from moist soil. The annual volume of water that naturally recharged to or discharged from the groundwater flow system in the study area during predevelopment conditions was estimated to be 1,010 acre-feet per year (acre-ft/yr). About 90 percent of this recharge originated as runoff from the Little San Bernardino and the Pinto Mountains to the south, and the remainder originated as runoff from the San Bernardino Mountains to the west. Evapotranspiration by phreatophytes near Mesquite Lake (dry) was the primary form of predevelopment groundwater discharge. From 1953 through 2007, approximately 139,400 acre-feet (acre-ft) of groundwater was pumped by the MCAGCC from the Surprise Spring subbasin.

A regional-scale numerical groundwater flow model was developed using MODFLOW–2000 for the Surprise Spring, Deadman, Mesquite, and Mainside subbasins. The aquifer system was simulated by using three model layers representing the upper, middle, and lower aquifers. Measured groundwater levels for predevelopment conditions (before 1953) and for the period 1953 through 2007 were used to calibrate the groundwater-flow model. The simulated steady-state (predevelopment) recharge was about 980 acre-ft/yr; about 90 percent of the recharge was in the Mesquite subbasin. Most of the simulated steady-state discharge occurred as evapotranspiration at the Mesquite Lake (dry). A total of about 145,450 acre-ft of groundwater was simulated as being pumped from the model domain during the transient simulation period (1953–2007); about 139,400 acre-ft of the total was extracted from the Surprise Spring subbasin. The transient simulation indicates that almost all of the groundwater pumped in the Surprise Spring subbasin comes from groundwater storage, which is consistent with the measured long-term declines in groundwater levels.

The calibrated groundwater model was used to evaluate the potential effects on water levels and aquifer conditions in the Surprise Spring, Deadman, Mesquite, and Mainside subbasins for water-management strategies being considered by the MCAGCC to meet the projected water demand at the base for 2008–2017. One of the main objectives of the water-management strategies is to reduce pumpage from the Surprise Spring subbasin. Reducing groundwater pumpage in the Surprise Spring subbasin by about 38 percent (about 1,345 acre-ft/yr) substantially decreased or reversed simulated hydraulic-head declines in the subbasin. Redistributing about 15 percent of the 2007 groundwater pumpage (about 550 acre-ft/yr) from the Surprise Spring to the Mainside subbasin resulted in more than 60 ft of simulated declines in hydraulic head in the Mainside subbasin by 2017; however, redistributing about 22 percent of the 2007 pumpage (about 800 acre-ft/yr) from the Surprise Spring subbasin to the Deadman subbasin resulted in 5–10 ft of simulated hydraulic-head decline in the Deadman subbasin. The water-management scenarios simulated for this study demonstrate how the calibrated regional model can be utilized to evaluate the hydrologic effects of a water-management strategy.

Introduction

The Marine Corps Air Ground Combat Center (MCAGCC) Twentynine Palms, California, is the largest air/ground combat training center of the U.S. Marine Corps, occupying about 932 square miles (mi^2) of the southern Mojave Desert. The MCAGCC overlies parts of the Surprise Spring, Deadman, Mesquite, and Mainside subbasins of the Morongo groundwater basin. Historically, the MCAGCC has relied on groundwater pumped primarily from the Surprise Spring subbasin to provide almost all of its water supply. Groundwater pumpage in the Surprise Spring subbasin has caused groundwater levels in the subbasin to decline by as much as 190 ft during 1953 through 2007. Groundwater from other groundwater subbasins within the boundaries of the MCAGCC in the Morongo groundwater basin (Deadman, Mainside, and Mesquite) contains high concentrations of fluoride, arsenic, and (or) dissolved solids (Riley and Worts, 1953 [2001]), making it unsuitable for potable uses without treatment. Future water demands associated with the MCAGCC and the surrounding civilian population will place additional demands on the Morongo groundwater basin. The loss of aquifer storage in the Surprise Spring subbasin combined with the poor water quality of adjacent subbasins highlights the need to thoroughly understand and effectively manage the available groundwater resources in the Twentynine Palms area.

Purpose and Scope

In 2002, the U.S. Geological Survey (USGS) began a cooperative study with the MCAGCC to improve the understanding of the geohydrology of the Surprise Spring, Deadman, Mesquite, and Mainside subbasins and to develop a regional groundwater-flow model of these subbasins to help manage the water resources of the region. This report (1) describes the geohydrology of the Twentynine Palms area, (2) provides an overview of regional groundwater-flow and water-quality conditions, (3) documents the development and calibration of a regional groundwater-flow model, and (4) demonstrates utilization of the model to assess water-management strategies being considered by water-resource planners at the MCAGCC.

A computerized Geographic Information System (GIS) was developed to store hydrological, geological, geophysical, and geochemical data collected and compiled for this study. The GIS facilitates the understanding of the groundwater-flow system and the conceptualization of the groundwater-flow model.

Description of the Study Area

The study area consists of the Surprise Spring, Deadman (southern part), Mesquite, and Mainside subbasins of the Morongo groundwater basin in the southern Mojave Desert, about 130 miles (mi) east of Los Angeles and 5 mi north of Twentynine Palms (fig. 1). The Morongo groundwater basin underlies a broad, eastward-sloping alluvial desert plain, completely surrounded by mountains and uplands. The basin is divided into 17 groundwater subbasins by multiple faults and an anticlinal structure known as the Transverse Arch (Riley and Worts, 1953[2001]). The study area covers about 220 mi^2 and is bounded by the Hildalgo Mountain to the north, the Bullion Mountains to the east, the Pinto Mountain Fault to the south, and the Emerson and the Copper Mountain Faults to the west (fig. 2). The southwestern part of the MCAGCC overlies the eastern part of the groundwater basin, which is the sole source of drinking water for the MCAGCC and surrounding communities.

The regional climate of the Twentynine Palms area is characterized by hot, arid summers and cool winters. Average annual precipitation ranges from about 8 inches (in.) in the mountainous areas to about 4 in. on the desert floor (Nishikawa and others, 2004). In the mountainous areas, approximately 50 percent of the annual precipitation falls during the winter months (January through March) and 10 percent falls during the summer (July through September). On the desert floor, approximately 30 percent of the annual precipitation falls during the winter and 44 percent falls during the summer (Nishikawa and others, 2004). The average potential evapotranspiration rate of the region is 66.5 in. per year (California Irrigation Management Information System, 2002). Most of the precipitation falling in the mountains and on the desert

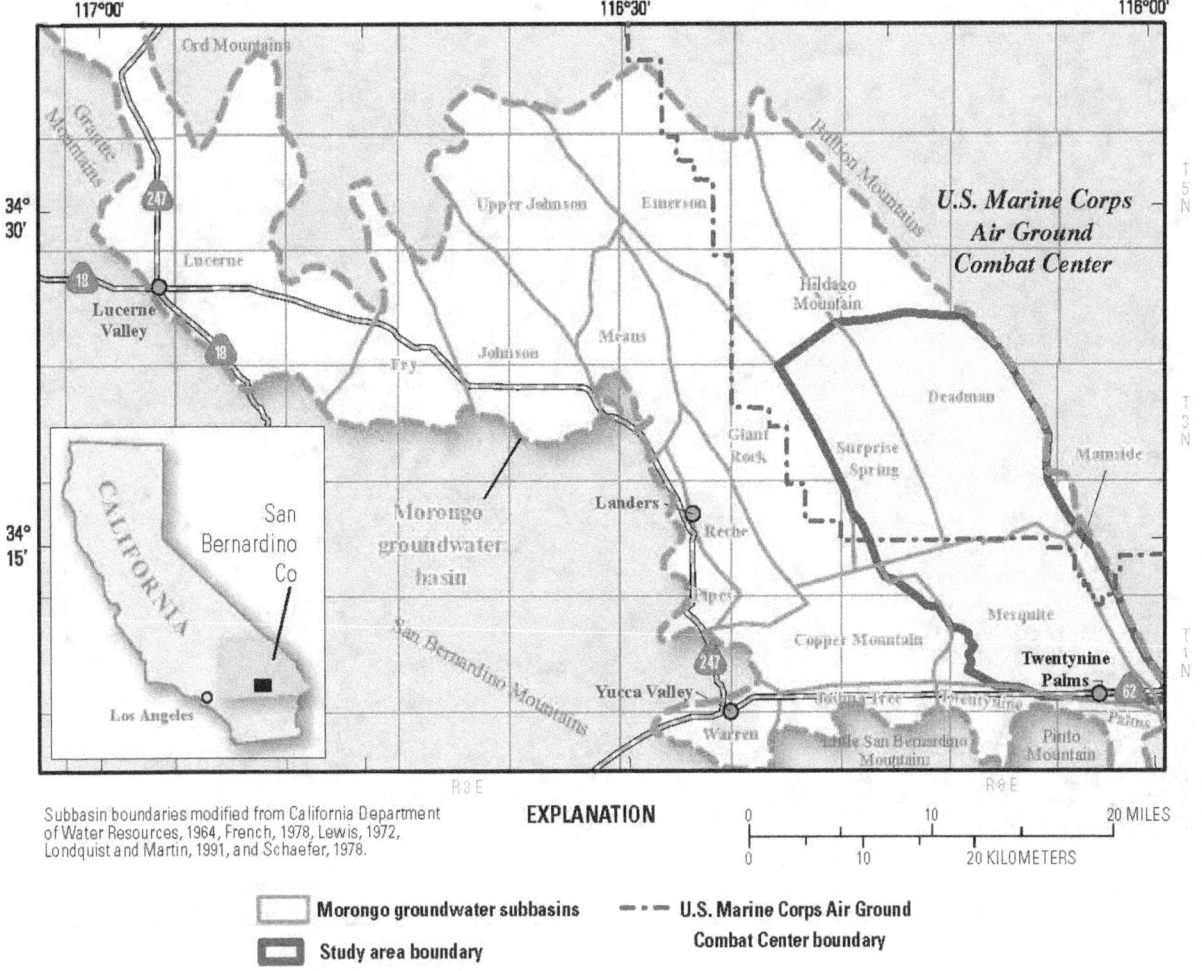

Figure 1.   Location of study area, and subbasins of the Morongo groundwater basin, California.

Subbasin boundaries modified from California Department
of Water Resources, 1964, French, 1978, Lewis, 1972,
Londquist and Martin, 1991, and Schaefer, 1978.

EXPLANATION

☐ Morongo groundwater subbasins

▭ Study area boundary

— ·· — U.S. Marine Corps Air Ground
Combat Center boundary

floor is lost through evapotranspiration; the remainder either seeps into the soil or remains on the surface to form overland flow when the rainfall is greater than the infiltration capacity of the soil. Streamflow in the study area is intermittent and occurs primarily as a result of high-intensity rainfall. Infiltration from streamflow is the major source of groundwater recharge (Riley and Worts, 1953 [2001]).

The study area receives surface-water runoff and recharge from the Emerson, Joshua Tree, Deadman, and Mesquite surface-water drainage basins (fig. 2). Each of these surface-water drainage basins contains a playa, or dry lake, at its lowest elevation (Emerson, Coyote, Deadman, and Mesquite Lakes). During periods of heavy runoff, water accumulates on the playas and forms shallow lakes. The Emerson surface-water drainage basin covers about 300 mi²; its headwaters are in the San Bernardino Mountains to the west and it discharges to Emerson Lake (dry). The Joshua Tree surface-water drainage basin covers about 260 mi²; its headwaters are in the

Little San Bernardino Mountains to the south. The Deadman surface-water drainage basin covers about 230 mi²; it drains the Bullion Mountains to the east and the surrounding desert floor. The Mesquite surface-water drainage basin covers about 178 mi²; its headwaters are in the Little San Bernardino and the Pinto Mountains.

Twentynine Palms and the main base area of the MCAGCC (fig. 2) are the principal areas of population within the study area. The total population at the beginning of 2006 was estimated to be about 27,500, and the projected growth rate is 2.1 percent per year (*http://www.ci.twentynine-palms. ca.us/City_Demographic_Data.63.0.html*). More than two-thirds of the population in the study area consists of military personnel and their family members, and the remainder consists of civilians who reside primarily in Twentynine Palms. The study area is essentially rural and undeveloped except for the main base area at MCAGCC and the city of Twentynine Palms.

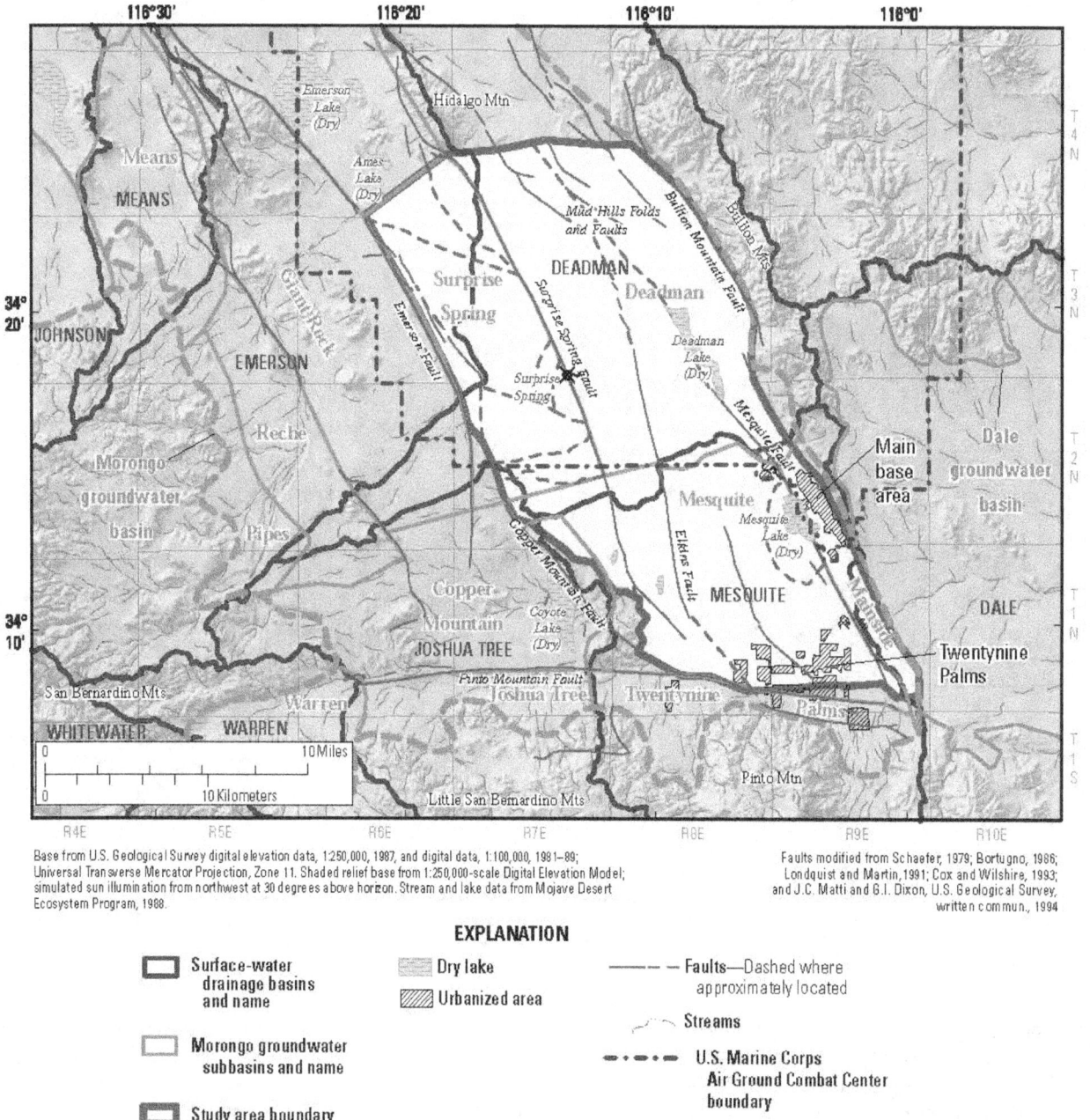

Base from U.S. Geological Survey digital elevation data, 1:250,000, 1987, and digital data, 1:100,000, 1981–89; Universal Transverse Mercator Projection, Zone 11. Shaded relief base from 1:250,000-scale Digital Elevation Model; simulated sun illumination from northwest at 30 degrees above horizon. Stream and lake data from Mojave Desert Ecosystem Program, 1988.

Faults modified from Schaefer, 1979; Bortugno, 1986; Londquist and Martin, 1991; Cox and Wilshire, 1993; and J.C. Matti and G.I. Dixon, U.S. Geological Survey, written commun., 1994.

EXPLANATION

☐ Surface-water drainage basins and name

☐ Morongo groundwater subbasins and name

☐ Study area boundary

▨ Dry lake

▨ Urbanized area

— — — Faults—Dashed where approximately located

⌒⌒ Streams

—·—·— U.S. Marine Corps Air Ground Combat Center boundary

Figure 2. Study area boundary, surface-water drainage area boundaries, and subbasins of the Morongo groundwater basin in the Twentynine Palms area, California.

Previous Investigations

Since the MCAGCC was established in 1950s, many investigations have been completed by the USGS on the geology and the groundwater resources of the base and surrounding area. These investigations include a geologic reconnaissance and test drilling program at the base (Riley and Worts, 1952[2001]), a geology and groundwater appraisal of the base and surrounding area (Riley and Worts, 1953[2001]), an evaluation of the groundwater resources of the base and surrounding area (Schaefer, 1978), a Bouger gravity anomaly map for the base and surrounding area (Moyle, 1984), an evaluation of the geohydrology and the potential for artificial recharge in the aquifers underlying the base (Akers, 1986), an evaluation of the geohydrology and the groundwater-flow simulation of the Surprise Spring subbasin (Londquist and Martin, 1991), preliminary geological maps of the base (Matti and Morton, 1994, 1995), regional water-table maps of the Morongo groundwater basin (Trayler and Koczot, 1995; Mendez and Christensen, 1997; Smith and Pimentel, 2000;

Smith, 2002; Smith and others, 2004; and Stamos and others, 2004, 2007), a gravity map of the Morongo groundwater basin (Roberts and others, 2002), an evaluation of the geohydrologic framework, recharge estimates, and groundwater flow in the Joshua tree area (Nishikawa and others, 2004), an evaluation of the source, movement, and age of ground water in the western part of the Mojave Desert (Izbicki, 2004; Izbicki and Michel, 2004), and an evaluation of the source and movement of helium in the eastern Morongo groundwater basin (Kulongoski and others, 2005).

In addition to USGS investigations, numerous geohydrologic reports have been completed by consultants and other government agencies in cooperation with the U.S. Department of the Navy. Principal studies include a reconnaissance of groundwater resources of the area (Wagner, 1952) a geotechnical study of the main base (Wahler & Associates, 1983), an investigation of geothermal resources (Trexler and others, 1984), and a compilation of information on groundwater and wells within the Twentynine Palms water district (Haley and Aldrich, 2001). Many reports were prepared for the Southwest Division Naval Facilities Engineering Command (Navy) under the Comprehensive Long-Term Environmental Action Navy (CLEAN) program. Monitoring wells constructed during this program were used to collect water-level and water-quality data as part of the current study.

Geohydrology

The geohydrology of the Surprise Spring, Deadman, Mesquite, and Mainside subbasins of the Morongo groundwater basin was defined by summarizing previously published research, completing a gravity survey of the region to estimate the depth-to-basement complex, and collecting and evaluating geologic and hydrologic data from new and existing wells.

Geology

The Morongo groundwater basin underlies a broad, eastward-sloping alluvial desert plain, completely surrounded by mountains and uplands. The mountain ranges and uplands consist of a nearly impermeable complex of igneous and metamorphic rocks of pre-Tertiary age (fig. 3). These rocks are the oldest geological materials in the study area and are referred to as basement complex in this report. The basement complex was uplifted millions of years ago, forming the mountains, and is buried by younger alluvial deposits throughout most of the Morongo groundwater basin. During the last several million years, the lowland valleys of the groundwater basin have been accumulating gravel, sand, silt and clay deposited mainly by streams flowing from the surrounding mountains. These alluvial deposits are augmented by wind-blown sand throughout the study area. Fine-grained silt and clay, as well as evaporative minerals such as gypsum and anhydrite, have been accumulating in the playa lakes (fig. 3) that periodically developed

in valley flatlands (Matti and Morton, 1995). Intermittently throughout their geologic history, these various geologic materials have been broken and warped by faults and folds.

Geologic Units

The geological materials in the study area are grouped into five stratigraphic units for this report (fig. 3). From the oldest to the youngest, these units are (1) basement complex of pre-Tertiary age (pTb), (2) older sedimentary deposits of Tertiary age (Miocene and Pliocene) (Ts), (3) volcanic rocks of Tertiary age (late to middle Pliocene) (Tv), (4) alluvial fan deposits of Tertiary-Quaternary age (late Pliocene to Pleistocene) (QTf), and (5) younger alluvium (Qa) and playa deposits (Qp) of Quaternary age (Pleistocene to Holocene). Data from geological maps, geological and geophysical logs of wells, and investigations of outcrops in the study area were used to define these stratigraphic units. Structural and stratigraphic relationships within the Morongo groundwater basin are presented in two east-west and two north-south geologic sections of the study area (fig. 4). The younger alluvium (Qa) and playa deposits (Qp) are not shown on the geologic sections because of negligible thickness of these deposits.

Basement Complex

The basement complex (pTb) forms the surrounding mountains and highlands (fig. 3) and underlies the Morongo groundwater basin (fig. 4). This unit consists predominately of plutonic intrusive rocks, including abundant Jurassic quartz monzonite. Although intensively weathered to clay or clayey sand locally near the contact with overlying sedimentary units, the crystalline basement rocks are relatively impermeable compared to the overlying Tertiary and Quaternary sediments and are not considered to be a water-bearing unit. Tertiary volcanic rocks (Tv) outcrop in the San Bernardino Mountains to the west and in the Bullion Mountains to the northeast of the study area. The Tertiary volcanic rocks generally are nonwater-bearing.

Water-Bearing Units

Tertiary and Quaternary sediments overlie the basement rocks throughout most of the study area (fig. 4). The thickness of these deposits varies significantly; the maximum thickness is close to 22,000 feet (ft) in the Deadman subbasin (Roberts and others, 2002). Riley and Worts (1953[2001]) defined the lithology of the Tertiary sediments from field reconnaissance of strata exposed in a southeast-plunging anticline in the Mud Hills (sec. 6, T. 3 N., R. 8 E., and secs. 35 and 36, T. 4 N., R. 7 E). The strata exposed in the Mud Hills are roughly 950 ft thick, as estimated from their dips and areal distribution (Brett Cox, U.S. Geological Survey, written commun., 2000). These deposits dip generally to the south beneath the northern part of the Deadman subbasin suggesting that a thickness of deposits roughly equivalent to that in the Mud Hills is present

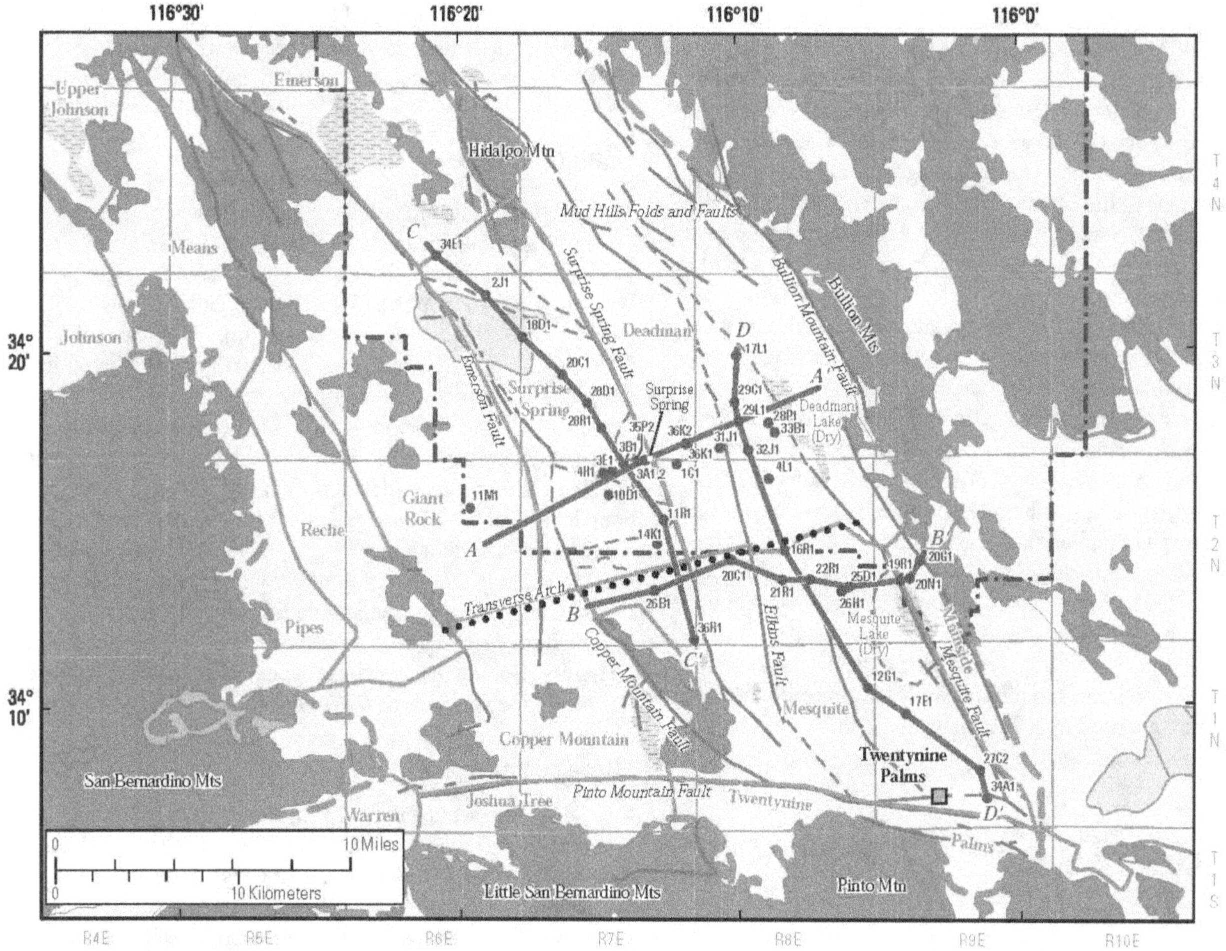

Base from U.S. Geological Survey digital data, 1:100,000, 1981–89;
Universal Transverse Mercator Projection, Zone 11

Faults modified from Schaefer, 1979; Bortugno, 1986;
Londquist and Martin,1991; Cox and Wilshire, 1993;
and J.C. Matti and G.I. Dixon, U.S. Geological Survey,
written commun., 1994

EXPLANATION

Stratigraphic units

Quaternary

- Qa Younger alluvium
- Qp Playa deposits

Quaternary/Tertiary

- QTf Alluvial fan deposits

Tertiary

- Ts Older sedimentary deposits
- Tv Volcanic rocks

pre-Tertiary

- pTb Basement complex

⊂▬⊃ ⊂▬⊃ **Morongo groundwater basin boundary**

──── **Morongo groundwater subbasin boundary**

──── **Dale groundwater basin boundary**

── ─ ─ **Faults**—Dashed where approximately
located.

••••• **Approximate axis of Transverse Arch**
(Riley and Worts, 1952)

──·─··─ **USMC Air Ground Combat Center boundary**

──── **Line of geologic section**—See figures 4A–D

²⁰C1 • **Wells included in geologic section and identifiers**

Figure 3.   Generalized geology of the Twentynine Palms area, California.

in the Surprise Spring and Deadman subbasins (Riley and Worts, 1953[2001]). Although there is lateral variation in the deposits, the major lithologic zones are recognizable across distances of as much as 5 mi in the Mud Hills suggesting that the general character of the section can be used to help describe the lithology of the study area. As part of this study,

the description of the Tertiary sediments exposed in the Mud Hills was used in conjunction with geologic logs and electric logs from wells in the Surprise Spring and Deadman subbasins to describe the lithology and stratigraphy of the Tertiary sediments in the study area (Brett Cox, U.S. Geological Survey, written commun., 2000).

A

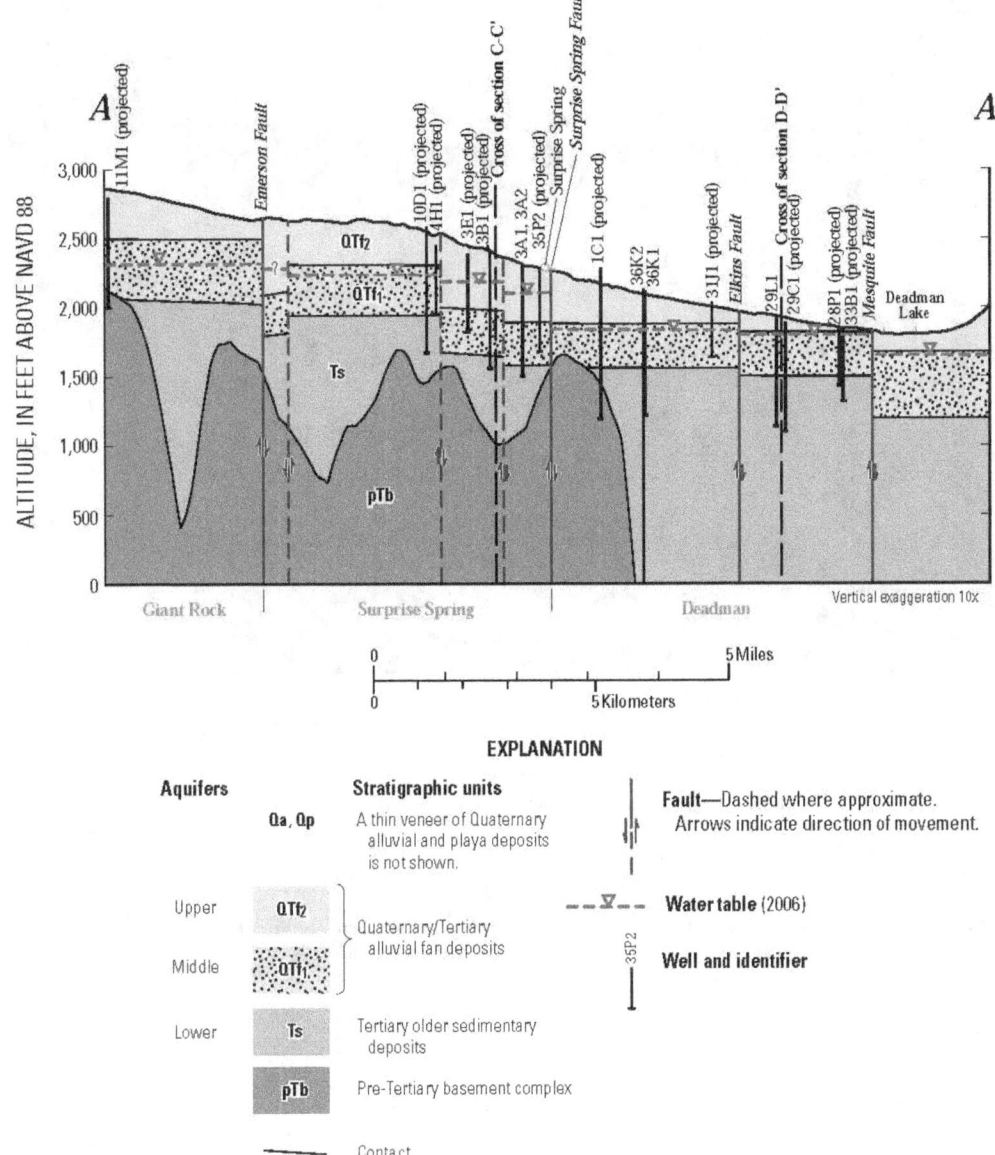

Figure 4. Generalized geologic cross section *A–A′*, *B–B′*, *C–C′* and *D–D′* of the Twentynine Palms area, California. Lines of geologic section are shown on fig. 3.

The older sedimentary deposits (Ts) consist of sand, gravel, and subordinate silt and clay that are commonly indurated with interstitial clay and calcium-carbonate cement. These deposits are characterized by low resistivity values in available electric logs from wells that penetrate the unit (Londquist and Martin, 1991). Low resistivity values on electric logs collected from wells that penetrate saturated unconsolidated deposits in the study area indicate either fine-grained deposits that do not yield water freely to wells or highly saline groundwater (Londquist and Martin, 1991). This geologic unit was deposited before the uplift of the San Bernardino Mountains; therefore, the major sources of deposits were alluvial fans originating from the mountains and upland areas in the southern Mojave Desert. Abundant rock fragments consisting of Jurassic quartz monzonite and Tertiary volcanic rocks derived from the Bullion Mountains and neighboring areas of the southern Mojave Desert are in this unit (Brett Cox, U.S. Geological Survey, written commun., 2000). The older sedimentary deposits are more consolidated with depth and yield a very limited amount of water to wells (Londquist and Martin, 1991).

B

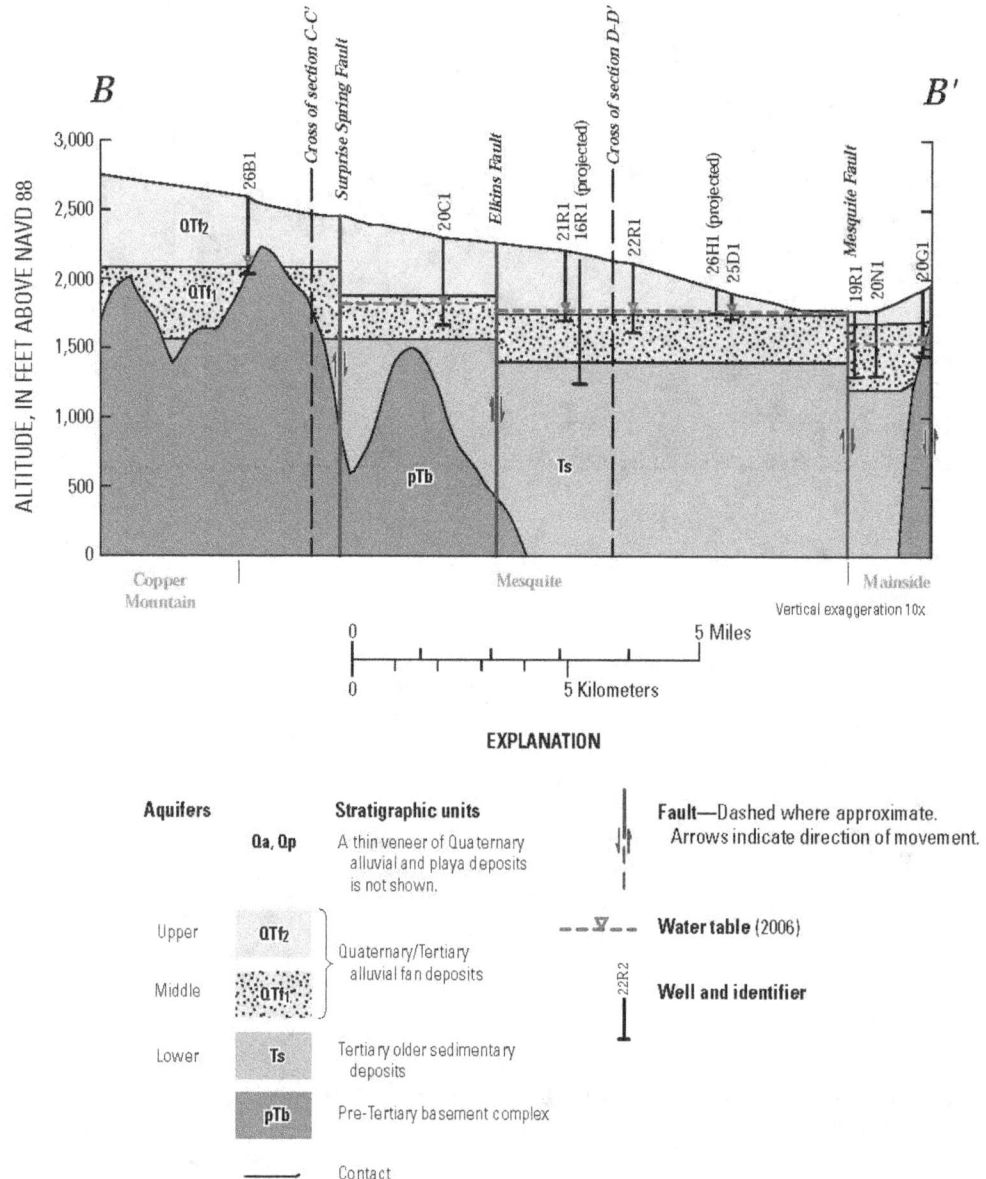

Figure 4. Continued.

The alluvial fan deposits (QTf) consist of varying amounts of gravel, sand, silt, and clay that originated predominantly from the eastern San Bernardino Mountains in the northern part of the study area and the Little San Bernardino and Pinto Mountains in the southern part of the study area. These deposits are characterized by high resistivity values in available electric logs from wells that penetrate the unit compared to the low resistivity values in the older sedimentary deposits (Londquist and Martin, 1991). High resistivity values on electric logs collected from wells that penetrate saturated unconsolidated deposits in the study area indicate coarse-grained deposits that yield water freely to wells (Londquist and Martin, 1991).

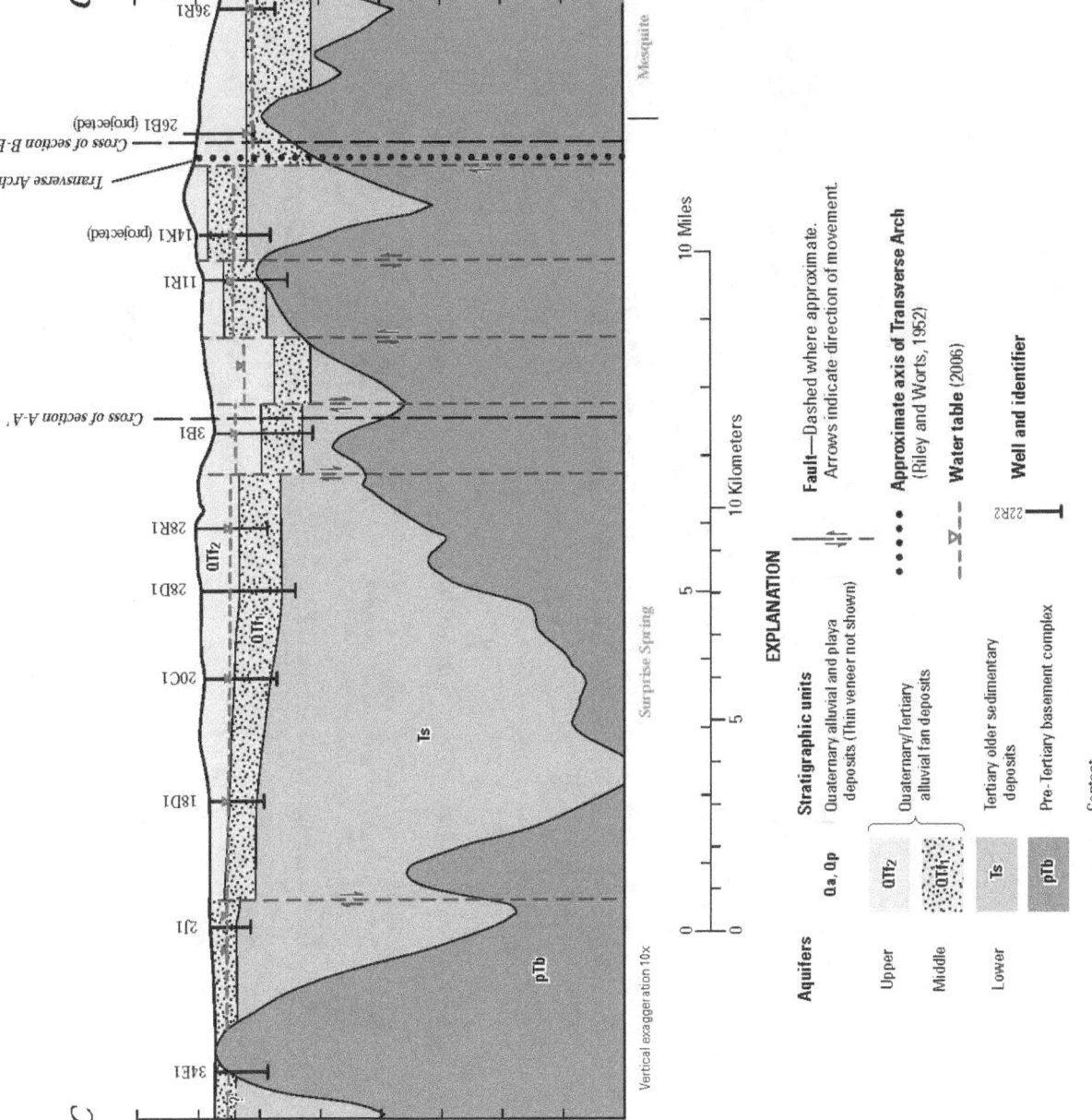

Figure 4.   Continued.

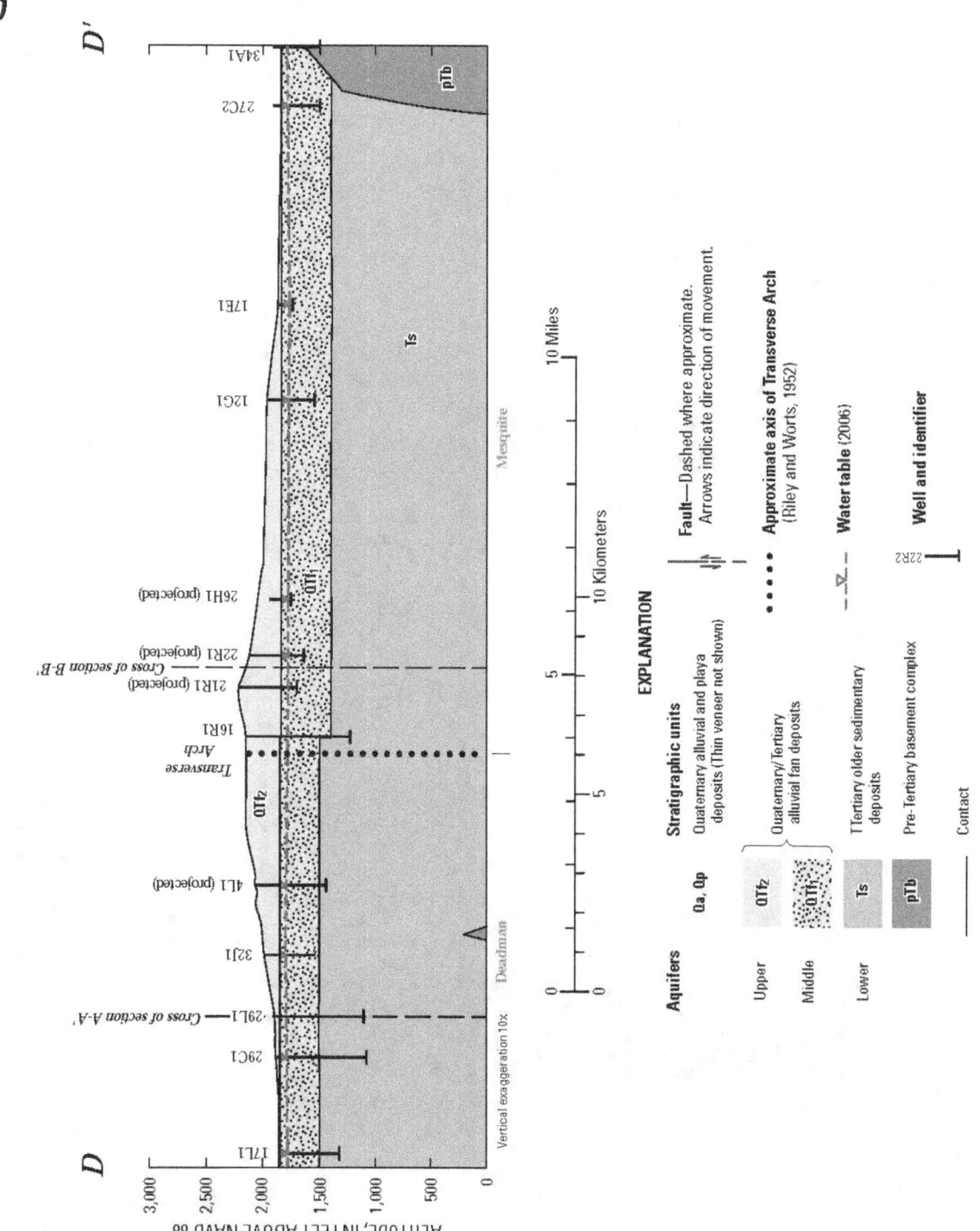

Figure 4. Continued.

The alluvial fan deposits were divided into an upper and a lower unit for this study on the basis of water-bearing characteristics. The lower unit (QTf_1) consists of silty sand and gravel, which are interbedded with moderate amounts of silt and clay that were deposited on the lower slopes of the alluvial fans. The lower unit is irregularly cemented with calcium carbonate and is moderately consolidated. The upper unit (QTf_2) consists of unconsolidated pebbly sand, pebble-cobble gravel, and minor silt and clay that were mainly stream deposits. In general, QTf_2 is more permeable than QTf_1 because of the predominance of the coarser-grained deposits and the lack of cementation. QTf_1 and QTf_2 are the two principal water-bearing units in the study area. The combined thickness of QTf_1 and QTf_2 ranges from less than 250 to more than 1,000 ft (fig. 4).

Young alluvium (Qa) and playa or dry-lakebed deposits (Qp) overlie the alluvial fan deposits as a thin veneer that is less than 50 ft thick throughout most of the study area. These deposits vary from poorly sorted sand and gravel in the alluvial fans to fine sand, silt, and clay in the playa (Londquist and Martin, 1991). In general, the young alluvium and playa deposits are above the water table and therefore are not an important water-bearing unit. The playa deposits are present at Emerson, Deadman, Coyote, and Mesquite Lakes (dry). Borehole data indicate that the playa deposits are about 50 ft thick at Deadman and Mesquite Lakes.

Depth to the Basement Complex

A gravity survey was done to understand the three-dimensional structure and estimate the depth to the basement complex (thickness of basin fill) in the study area (Roberts and others, 2002). The area of the gravity survey coincides with an earlier gravity survey of the Twentynine Palms area by Moyle (1984); however, the number of gravity stations is more than double the number used in the earlier study, improving the definition of the gravity field. Estimating the depth to the basement complex by using a gravity survey requires knowledge of the residual gravity field of the exposed geology and knowledge of the vertical density variation within the basin deposits. Data from boreholes that penetrate the surface of the basement complex, and any other geophysical data that provide constraints on the thickness of the basin fill, also are used.

The thickness of the alluvial deposits that form the groundwater basin was estimated by using the method developed by Jachens and Moring (1990), modified slightly to include constraints at points where the thickness of the basin fill is known from direct observations in drill holes. This method partitions the residual gravity field into two components—the component caused by density variations within the basement complex (the basement gravity field) and the component caused by the low-density basin fill that forms the groundwater basin (the "groundwater basin gravity anomaly"). Once the gravity data have been partitioned, the groundwater basin gravity anomaly can be modeled to yield a thickness of

the basin fill throughout the study area. For detailed information regarding the gravity survey and analysis, the reader is referred to Roberts and others (2002).

The results of the gravity modeling indicate that two basins beneath the Deadman and the Mesquite Lakes (dry) are more than 16,000 ft deep (fig. 5). No wells penetrate the entire thickness of basin fill near these deep depressions; therefore, the depth to the basement complex may be under or overestimated in these areas (Roberts and others, 2002). These deep basins are likely strike-slip extensional basins caused by a right slip across a right step from the Mesquite Fault to the Bullion Mountain Fault (Roberts and others, 2002).

The depth to the basement complex at the Surprise Spring subbasin is relatively shallow: the maximum depth is less than 4,300 ft. The estimated depth decreases to less than 400 ft at the southern end of Surprise Spring subbasin and the northwestern part of Mesquite subbasin near Copper Mountain. This rise of the basement complex is believed to be related to the east-west anticlinal structure referred to as the Transverse Arch.

A three-dimensional visualization of the stratigraphic units in the study area was created to illustrate the thickness and spatial variation of the stratigraphic units in the Surprise Spring, Deadman, and Mesquite subbasins (fig. 6). Deadman and Mesquite subbasins are significantly deeper than the Surprise Spring subbasin. The older sedimentary deposits (Ts) compose most of the sediments in the subbasins, with the alluvial fan deposits (QTf_2 and QTf_1) and the younger alluvium (Qa) and playa deposits (Qp) forming a thin crust overlying the older deposits. The Ts unit yields a small amount of water to wells. Despite the tremendous thickness of the sedimentary deposits in the study area, most of the area has limited groundwater resources because of the relatively thin layer of saturated alluvial fan deposits (QTf_2 and QTf_1) that yield water freely to wells.

Geologic Structure

The study area is dominated by extensive faulting and moderate to intense folding that have displaced or deformed the pre-Tertiary basement complex as well as Tertiary and Quaternary deposits (figs. 3, 4). These faults are primarily right-lateral strike-slip faults, mostly trending northwest to southeast. Knowledge of these faults and folds is important because geological structures often are barriers to the lateral movement of groundwater flow. The barrier effect of faults is caused by the low permeability of the fault zone resulting from the compaction and extreme deformation of the water-bearing deposits adjacent to the faults, and by lateral juxtaposition of high- and low-permeability units. Cementation of the fault zone by the deposition of minerals from rising groundwater also can contribute to reducing fault-zone permeability.

Geologic maps and water-level data indicate that the following geological structures within the study area are groundwater-flow barriers:

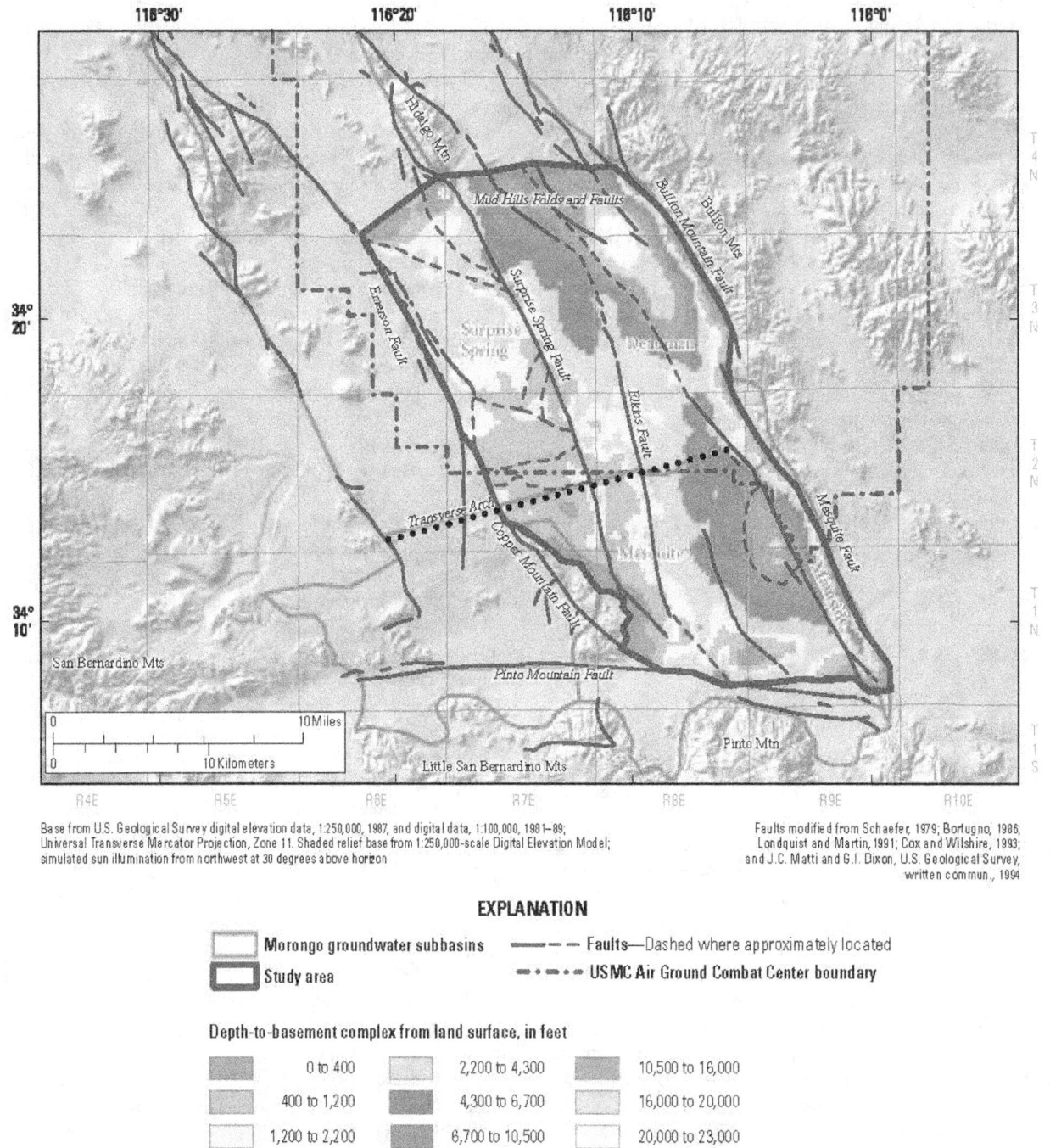

Base from U.S. Geological Survey digital elevation data, 1:250,000, 1987, and digital data, 1:100,000, 1981–89;
Universal Transverse Mercator Projection, Zone 11. Shaded relief base from 1:250,000-scale Digital Elevation Model;
simulated sun illumination from northwest at 30 degrees above horizon

Faults modified from Schaefer, 1979; Bortugno, 1986;
Londquist and Martin, 1991; Cox and Wilshire, 1993;
and J.C. Matti and G.I. Dixon, U.S. Geological Survey,
written commun., 1994

EXPLANATION

▢ Morongo groundwater subbasins   — — — Faults—Dashed where approximately located

▢ Study area   — ·· — ·· — USMC Air Ground Combat Center boundary

Depth-to-basement complex from land surface, in feet

0 to 400	2,200 to 4,300	10,500 to 16,000
400 to 1,200	4,300 to 6,700	16,000 to 20,000
1,200 to 2,200	6,700 to 10,500	20,000 to 23,000

Figure 5.   Depth to the basement complex, derived from gravity data for the Twentynine Palms area, California.

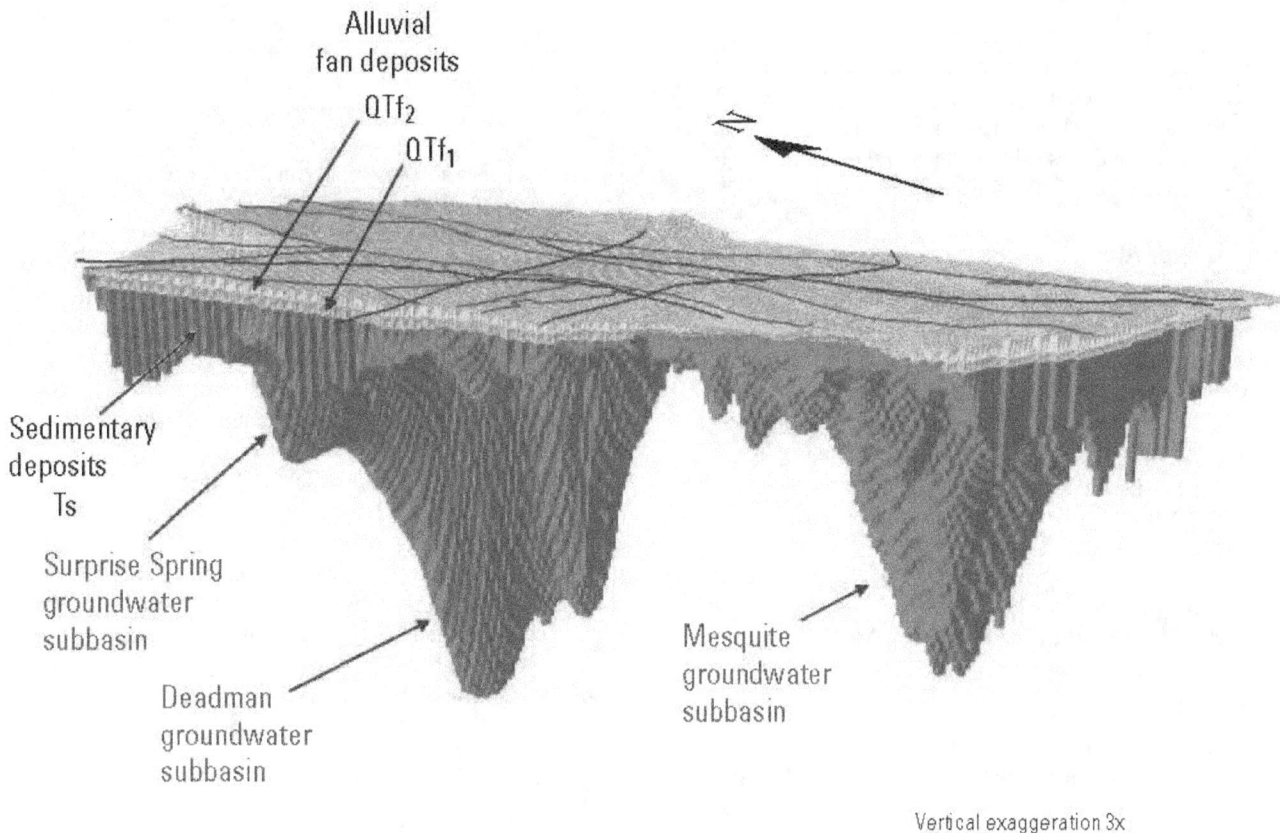

Alluvial
fan deposits
QTf₂
QTf₁

Sedimentary
deposits
Ts

Surprise Spring
groundwater
subbasin

Deadman
groundwater
subbasin

Mesquite
groundwater
subbasin

Vertical exaggeration 3x

Figure 6.   Three-dimensional view of thickness of sedimentary deposits in the Twentynine Palms area, California.

Bullion Mountain Fault

The Bullion Mountain Fault (fig. 3) is an eastward-tilted fault block, up-thrown by several normal faults along its extremely steep western face. The faulting has juxtaposed the nonwater-bearing basement complex on the east side of the fault against water-bearing sedimentary deposits on the west side of the fault. The Bullion Mountain Fault forms the eastern boundary of the Deadman and the Mainside subbasins.

Mesquite Fault

The Mesquite Fault is the northwest-southeast trending fault that separates the Mesquite and the Mainside subbasins (fig. 3). The barrier effect of the fault is most striking in the Mesquite Lake (dry) area, where a heavily vegetated area is found to the west of the fault, where the water table is at or near land surface; whereas, the area east of the fault is nearly barren and the depth to water exceeds 200 feet (Stamos and others, 2006). These sharply contrasting conditions indicate the effectiveness of the Mesquite Fault as a groundwater barrier.

Surprise Spring Fault

The Surprise Spring Fault is a northwest-southeast trend-ing fault that separates the Surprise Spring and the Deadman subbasins (fig. 3). The fault is an effective groundwater bar-rier, as evidenced by water levels that are about 400 ft lower in the Deadman subbasin than in the Surprise Spring subbasin (Londquist and Martin, 1991). Prior to groundwater pumping in the Surprise Spring subbasin, the barrier effect of the fault caused groundwater levels to be at or near land surface in the southern part of the subbasin , as evidenced by the Surprise Spring and mesquite trees on the west side of the fault.

Emerson Fault

The Emerson Fault is a northwest-southeast trending fault that forms the western boundary of the Surprise Spring subbasin (fig. 3). The Emerson Fault was formed by uplift of the basement complex along the western side of the fault, but it has not disrupted the younger alluvium deposited by Pipes Wash. The Emerson Fault is a barrier to flow, as evidenced by water levels that are as much as 100 ft higher in the Emerson subbasin than in the Surprise Spring subbasin (Stamos and others, 2006).

Pinto Mountain Fault

The Pinto Mountain Fault is an east-west trending fault that separates the Twentynine Palms and the Mesquite sub-basins (fig. 3). The Pinto Mountain Fault is a barrier to flow, as evidenced by water levels that are more than 100 ft higher in the Twentynine Palms subbasin than in the Mesquite subbasin (Stamos and others, 2006).

Transverse Arch

The Transverse Arch is an east-west trending anticline that forms a topographic high extending across the center of the study area (fig. 3). Gravity data compiled for this study (Roberts and others, 2002) indicate that the basement complex is within 500 feet of land surface along the alignment of the arch south of the Surprise Spring subbasin (fig. 5). The structure forms a groundwater barrier at the west end of the study area as evidenced by water levels that are about 200 ft higher north of the arch than south of the arch (Stamos and others, 2006). The Transverse Arch and related structures form the southern boundary of the Surprise Spring subbasin and separates the Deadman and Mesquite subbasins (fig. 3).

Other Faults

Londquist and Martin (1991) identified additional faults as partial groundwater-flow barriers in Surprise Spring sub-basin. The effects of these unnamed faults were manifested by abrupt water-level changes across the faults as ground-water use in the subbasin increased. Other faults that might be partial groundwater-flow barriers are the Elkins Fault and two unnamed faults east of the Elkins Fault in the Mesquite subbasin (fig. 3). Topographic and aeromagnetic data (Robert Jachens, U.S. Geological Survey, written commun., 2003) suggest that there may be many more unmapped faults in the study area, and the effect of these faults on the regional groundwater flow system, especially during increased ground-water withdrawals, is currently unknown.

Groundwater Hydrology

Definition of the Aquifer System

The water-bearing deposits in the study area comprise the Quaternary/Tertiary alluvial fan deposits (QTf$_2$ and QTf$_1$) and the Tertiary older sedimentary deposits (Ts) (figs. 3, 4). The Quaternary alluvial and playa deposits are unsaturated throughout most of the study area. Interpretations from litho-logic and downhole geophysical logs were used to identify two aquifers (referred to as the upper and the middle aquifers) in the Quaternary/Tertiary alluvial fan deposits and a single aquifer (referred to as the lower aquifer) in the Tertiary older sedimentary deposits. The base of the aquifer system consists of the pre-Tertiary basement complex.

The upper aquifer consists of stratigraphic unit QTf$_2$ (fig. 4) and is unconfined. This aquifer is dominated by sand and gravel, which is highly permeable and yields a large quan-tity of water to wells where it is saturated. Except for parts of the Surprise Spring and Mesquite subbasins, the upper aquifer is unsaturated (fig. 4). In 2002, the maximum saturated thick-ness of the upper aquifer was about 250 ft, near the Surprise Spring (fig. 4A).

The middle aquifer consists of stratigraphic unit QTf$_1$. It is dominated by sand, silt, and clay, and is less permeable than the upper aquifer. The thickness of this aquifer ranges from about 100 ft at the north part of the Surprise Spring subbasin to almost 500 ft east of the Mesquite Fault in the Deadman subbasin (fig. 4A). The middle aquifer is confined by the over-lying upper aquifer where the upper aquifer is saturated.

The lower aquifer consists of stratigraphic unit Ts. It contains poorly-sorted sands, gravel, silt, and clay that become more consolidated with depth (Londquist and Martin, 1991). The overall permeability of this aquifer is low. Its thickness varies greatly within the study area, from less than 100 ft near the Transverse Arch in the western part of Mesquite subbasin (fig. 4B) to more than 16,000 ft beneath the Deadman and Mesquite Lakes (dry) (figs. 4, 5). The lower aquifer is confined by the overlying middle aquifer throughout the study area.

Aquifer Properties

Aquifer properties, including transmissivity, hydraulic conductivity, and storage coefficient, affect the rate at which water moves through the aquifer, the amount of water in storage, and the rate and areal extent of water-level declines caused by groundwater development. Transmissivity is a mea-sure of the ability of an aquifer to transmit water, and hydrau-lic conductivity is the capacity of a rock or unconsolidated material to transmit water. The transmissivity of an aquifer is equal to the hydraulic conductivity multiplied by the aquifer thickness. The storage coefficient of an aquifer is the volume of water released from or taken into storage per unit of surface area per unit change in head (Lohman, 1972).

The aquifer properties of the Twentynine Palms area were estimated for this study from well logs, specific-capacity tests, and published data. For the purposes of this study, the four groundwater subbasins being investigated (Surprise Spring, Deadman, Mesquite, and Mainside) were subdivided into hydrogeologic zones (fig. 7). Adjacent hydrogeologic zones are separated by a fault that is a barrier or a partial barrier to groundwater flow. Water-level altitudes and aquifer properties are similar within each zone of each aquifer.

Transmissivity and Hydraulic Conductivity

The transmissivity and hydraulic conductivity of the upper and middle aquifers were estimated from specific-capacity data. Specific capacity is the yield of a well per unit of drawdown and is a function of the transmissivity of

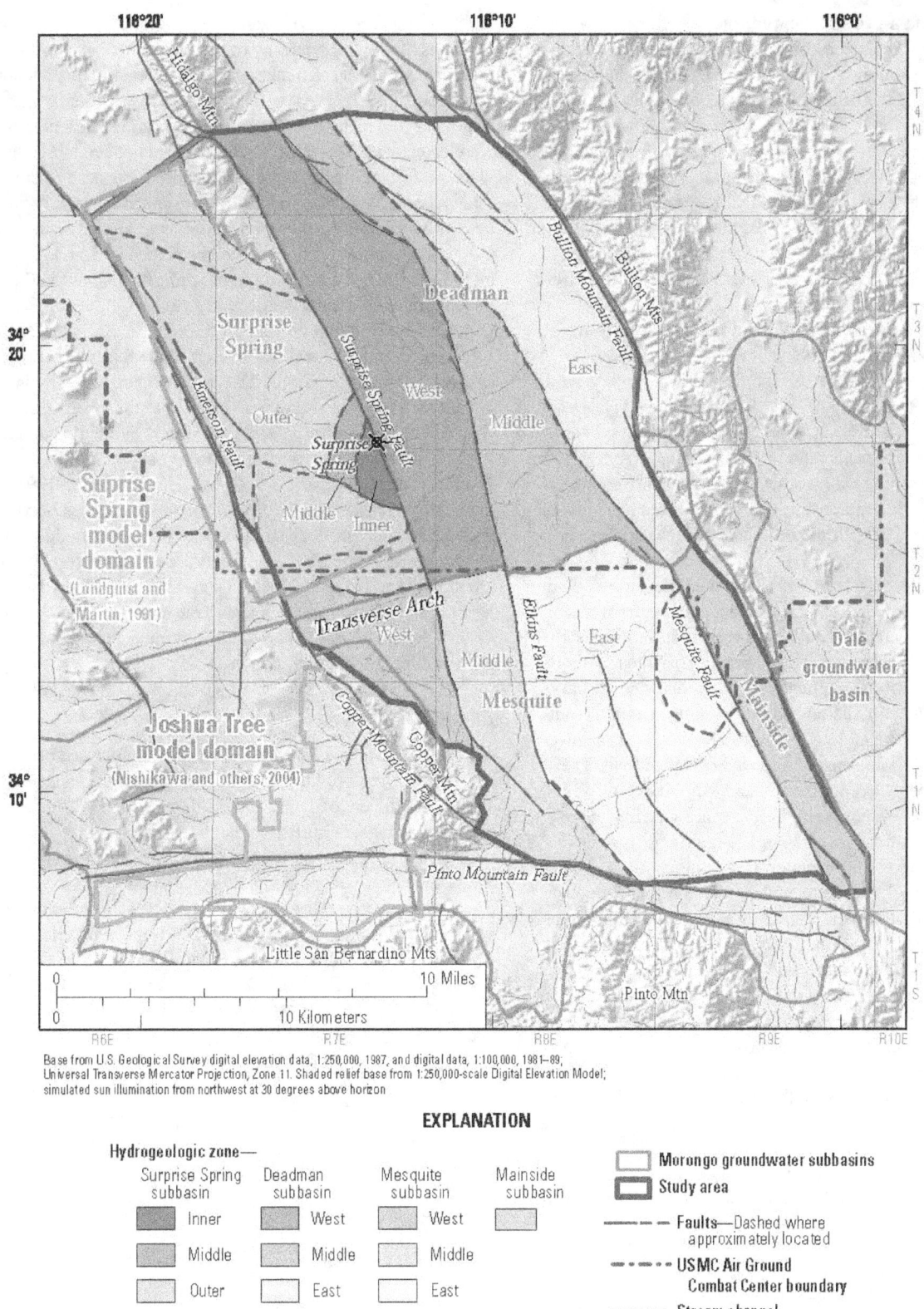

Figure 7.   Groundwater subbasins and hydrogeologic zones in the Twentynine Palms area, California.

the aquifer and aspects of the well, such as efficiency and borehole storage. Thomasson and others (1960) reported that for unconfined valley-fill deposits in the Sacramento Valley, California, the specific capacity in units of gallons per minute per foot multiplied by 230 approximates the transmissivity of the aquifer in units of feet squared per day. This relation between specific capacity and transmissivity was assumed to be representative of the upper and middle aquifers in the study area. The hydraulic conductivity of the upper and the middle aquifers was estimated by dividing the estimated transmissivity of the aquifers by the saturated thickness of each aquifer (table 1).

Specific-capacity data for wells in the study area were compiled from previous publications (Riley and Worts, 1953[2001]; Charles Kaehler, U.S. Geological Survey, written commun., 1991; Londquist and Martin, 1991; and Haley and Aldrich, 2001) (table 1). A total of 32 specific-capacity tests were compiled for wells ranging in depth from 120 to 1,250 ft. The locations of the wells presented in table 1 are shown on figure 8. Specific capacity of these wells ranged from less than 1 to more than 100 gallons per minute per foot ((gal/min)/ft) of drawdown.

Londquist and Martin (1991) divided the Surprise Spring subbasin into three zones (inner, middle, and outer zones) (fig. 7), each containing wells with similar specific-capacity values (table 1). The inner zone consists of an area west of the Surprise Spring fault, northeast of an unnamed northwest-southeast trending fault, and east of an unnamed north-south trending fault. The middle zone consists of an area northwest of the inner zone and east of a second unnamed fault. The outer zone consists of the remainder of the subbasin. Wells in the inner zone have the highest average specific-capacity values, ranging from about 74 to 100 (gal/min)/ft; wells in the middle zone have intermediate values, ranging from about 29 to 34 (gal/min)/ft; and wells in the outer zone have the lowest values, ranging from about 17 to 25 (gal/min)/ft. The decrease of specific-capacity values from the inner to the outer zone is a result of the reduced saturated thickness of the permeable upper aquifer (QTf$_2$) in the middle and outer zones of the Surprise Spring subbasin compared to the inner zone (figs. 4A,C). The transmissivity values estimated from the specific-capacity data range from to 3,900 feet squared per day (ft^2/d) in the outer zone to 23,000 ft^2/d in the inner zone (table 1). The estimated hydraulic conductivity values range from 12.3 feet per day (ft/d) in the outer zone to 40.5 ft/d in the inner zone.

The saturated part of layer QTf$_2$ is thin (less than 50 ft), or the unit is unsaturated, in Mesquite, Deadman, and Mainside subbasins (fig. 4). Therefore, wells in these subbasins are perforated mostly in the middle and the lower aquifers. The specific-capacity values of wells in Mesquite, Deadman, and Mainside subbasins are near the low end of the values of wells in the Surprise Spring subbasin (table 1). The geometric means of the estimated transmissivity and hydraulic conductivity values in these subbasins were 1,895 ft^2/d and 11.2 ft/d, respectively. The geometric mean is used instead of the arithmetic mean because the range in estimated transmissivity

and hydraulic conductivity values is large. A geometric mean indicates the central tendency or typical value of a set of numbers. Unlike an arithmetic mean, a geometric mean tends to dampen the effect of very high or very low values, which might bias the mean if a straight average (arithmetic mean) were calculated. Specific-capacity data are not available for wells perforated solely in the lower aquifer. Because of the relatively high degree of consolidation, specific-capacity values for wells perforated in this aquifer are expected to be low.

In a groundwater-flow model developed by Londquist and Martin (1991) for the Surprise Spring subbasin (fig. 7), the calibrated hydraulic conductivity values for the alluvial fan deposits (upper and middle aquifers) were 33, 25, and 22 ft/d for the inner, middle, and outer zones in the Surprise Spring subbasin, respectively (table 2). A uniform horizontal hydraulic conductivity of 1 ft/day was assigned to the lower aquifer (Londquist and Martin, 1991). The vertical hydraulic conductivity of the aquifers was assumed to be one-tenth of the horizontal hydraulic conductivity (Londquist and Martin, 1991). In the nearby Joshua Tree area (fig. 7), model calibrated horizontal hydraulic conductivity values of the upper, middle and lower aquifers were 60, 5, and 0.5 ft/d, respectively (table 2) (Nishikawa and others, 2004). The vertical hydraulic conductivity of the aquifers in the Joshua Tree area was assumed to be one-hundredth of the horizontal hydraulic conductivity of each aquifer (Nishikawa and others, 2004).

Storage Coefficient

The storage coefficient of a confined aquifer can be estimated by multiplying the aquifer thickness by a specific-storage coefficient of 1×10^{-6} ft^{-1} (Lohman, 1972). This provides a reasonable estimate for confined aquifers that have not undergone significant compaction. For unconfined aquifers, the storage coefficient is virtually equal to the specific yield. The specific yield of an aquifer is the ratio of the volume of water that will drain freely, by gravity, from the material to the total volume of the material. It will always be less than the porosity of the aquifer materials. Specific yields of the upper and middle aquifers in the Surprise Spring subbasin were estimated to be 16 percent for both the outer and the middle zones and 25 percent for the inner zone (Londquist and Martin, 1991). The value of specific yield estimated for the Joshua Tree aquifer was 20 percent (Nishikawa and others, 2004). The specific yield of the lower aquifer in the Surprise Spring subbasin was assumed to be 5 percent (Londquist and Martin, 1991).

Natural Recharge and Discharge

The principal source of recharge to the study area was groundwater underflow across the western and southern boundaries, originating as runoff in the surrounding mountains (fig. 9). Recharge from the direct infiltration of streamflow at land surface is small compared to groundwater underflow and occurs only during large storm events. Recharge from direct

Table 1. Estimated transmissivity and hydraulic conductivity values based on specific capacity of wells in the Twentynine Palms area, California.

[See *figure 8* for locations of wells. **Abbreviations:** ft, foot; (gal/min)/ft, gallon per minute per foot; ft²/d, foot squared per day; ft/d, foot per day; NA, not available]

State well number	Well depth (ft)	Land surface altitude (ft)	Saturated thickness (ft)	Water level altitude (ft)	Length of perforation in aquifer (ft)			Specific capacity [(gal/min)/ft]	Estimated transmissivity (ft²/d)	Estimated hydraulic conductivity (ft/d)
					Upper	Middle	Lower			
Surprise Spring Subbasin										
Inner zone										
2N/7E-02D1	532	2,292	429	2,189	162	120	0	[1]74.3	17,089	39.8
2N/7E-03A1	550	2,303	495	2,248	213	127	0	[1]87.2	20,056	40.5
3N/7E-35P2	593	2,247	593	2,247	174	226	0	[1]100.2	23,046	38.9
Middle zone										
2N/7E-03B1	700	2,358	596	2,253	118	312	0	[1]29.1	6,693	11.2
2N/7E-03E1	510	2,388	357	2,234	158	102	0	[1]33.7	7,751	21.7
Outer zone										
3N/7E-28R1	600	2,524	321	2,245	0	190	0	[1]17.1	3,933	12.3
3N/7E-29R1	600	2,522	329	2,252	0	190	0	[1]18.4	4,232	12.8
3N/7E-32J1	600	2,551	295	2,246	0	190	0	[1]25.3	5,819	19.8
Mesquite Subbasin										
Middle zone										
1N/8E-05G1	371	2,190	61	1,880	0	60	0	[2]6.9	1,586	26.0
1N/8E-05H1	357	2,170	62	1,875	0	60	0	[2]21.4	4,915	79.3
East zone										
1N/8E-02C1	324	2,050	71	1,797	46	14	0	[2]6.2	1,426	20.1
1N/8E-12J1	250	1,950	63	1,763	20	40	0	[2]133.3	30,667	486.8
1N/9E-06E3	126	1,780	66	1,720	0	10	0	[2]2.1	30	.4
1N/9E-07D1	310	1,900	160	1,750	0	80	0	[2]33.5	7,712	48.2
1N/9E-07F1	200	1,900	71	1,771	0	40	0	[2]2.0	460	6.5
1N/9E-08R7	190	1,850	142	1,802	53	80	0	[2]30.0	6,900	48.6
1N/9E-16Q1	304	1,800	272	1,768	0	156	0	[2]3.3	759	2.8
1N/9E-21H1	1,250	1,840	1,200	1,790	0	85	570	[2]54.5	12,545	10.5
1N/9E-22D4	120	1,800	82	1,762	0	40	0	[2]2.5	575	7.0
1N/9E-04N1	495	1,789	483	1,777	37	340	106	[3]6.2	1,426	3.0
1N/9E-05G1	428	1,781	423	1,776	36	340	47	[3]33.0	7,590	18.0
2N/8E-24H1	320	1,859	240	1,778	38	202	0	[3]1.9	437	1.8
2N9E-19D4	519	1,862	430	1,773	0	203	57	[4]10.0	2,300	5.3
2N/9E-19D4	519	1,862	430	1,773	0	203	57	NA	18,70 4	4.3

Table 1. Estimated transmissivity and hydraulic conductivity values based on specific capacity of wells in the Twentynine Palms area, California.—Continued

[See *figure 8* for locations of wells. **Abbreviations:** ft, foot; (gal/min)/ft, gallon per minute per foot; ft²/d, foot squared per day; ft/d, foot per day]

State well number	Well depth (ft)	Land surface altitude (ft)	Saturated thickness (ft)	Water level altitude (ft)	Length of perforation in aquifer (ft)			Specific capacity [(gal/min)/ft]	Estimated transmissivity (ft²/d)	Estimated hydraulic conductivity (ft/d)
					Upper	Middle	Lower			
Deadman Subbasin										
West zone										
3N/7E–13N1	487	2,022	298	1,833	0	273	25	[3]11.2	2,576	8.7
Middle zone										
3N/8E–17L1	456	1,853	409	1,806	0	105	103	[3]1.6	368	.9
3N/8E–29C1	800	1,893	713	1,806	0	0	184	[3]8.6	1,978	2.8
3N/8E–29L1	590	1,908	488	1,806	0	138	182	[3]31.5	7,245	14.8
3N/8E–33B1	512	1,848	469	1,805	0	212	164	[3]21.6	4,968	10.6
3N/8E–34D1	396	1,826	372	1,802	0	140	70	[3]13.5	3,105	8.4
Mainside Subbasin										
1N9E–10M1	332	1,810	60	1,538	0	60	0	[2]5.6	1,288	21.5
2N/9E–20M1	390	1,802	136	1,548	0	140	0	[4]40.0	9,200	67.6

[1] Londquist and Martin, 1991

[2] Haley and Aldrich, 2001

[3] Riley and Worts, 1953[2001]

[4] Charles Kaehler, U.S. Geological Survey, written commun., 1991.

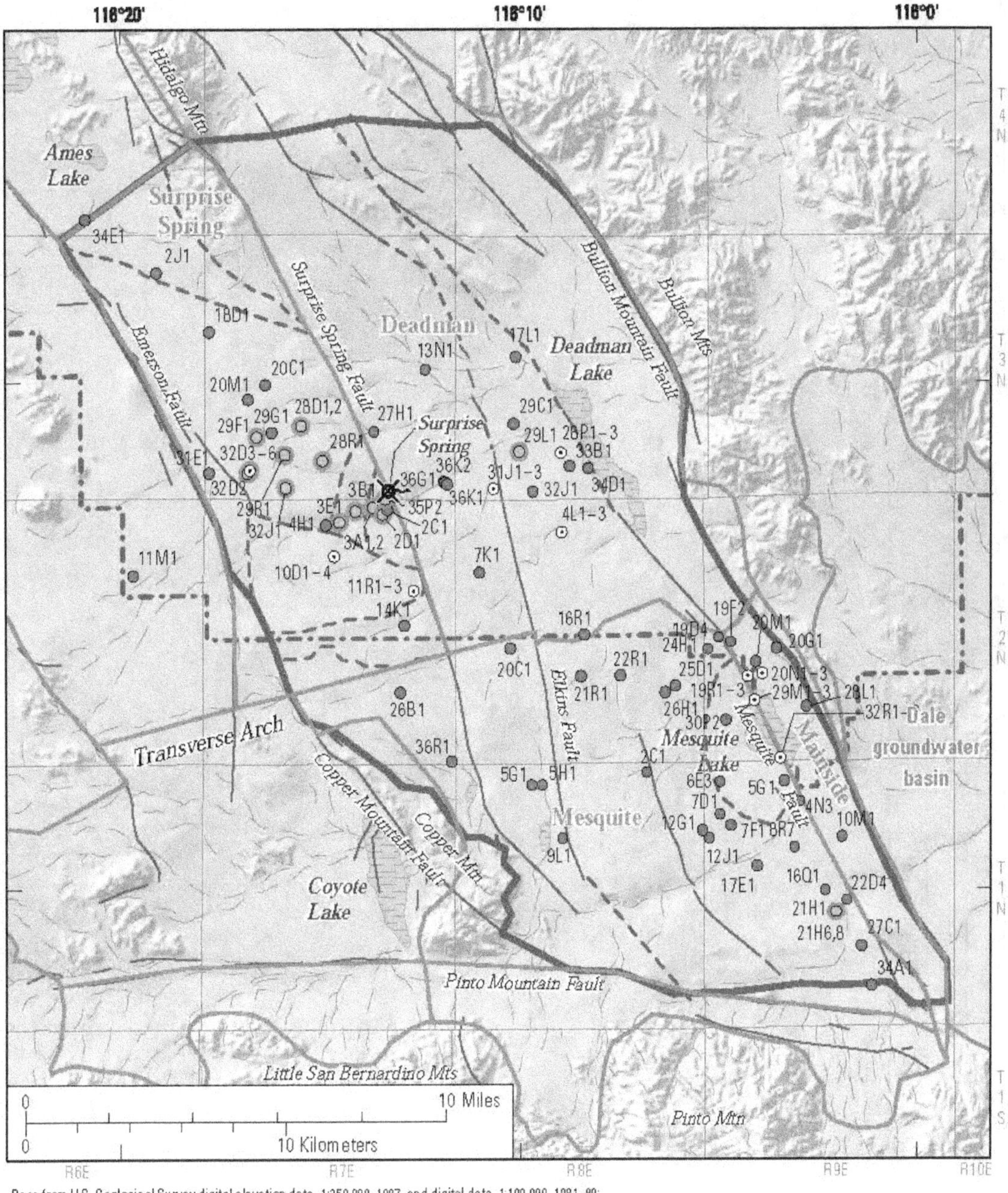

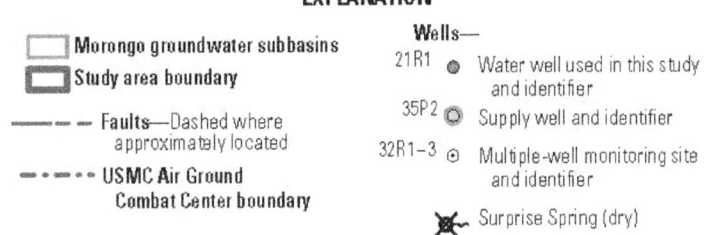

Figure 8. Location of selected wells in the Twentynine Palms area, California.

Table 2. Hydraulic-conductivity, specific-storage, and specific-yield values estimated for previous models of the Twentynine Palms area, California.

[The Surprise Spring model has two layers: layer 1 represents the upper and middle aquifers and layer 2 represents the lower aquifer. The Joshua Tree model has three layers: layer 1 represents the upper aquifer, layer 2 represents the middle aquifer, and layer 3 represents the lower aquifer. ft/d, foot per day; ft, foot; ft³, foot cubed; NA, not available]

Model aquifer	Zone	Hydraulic conductivity (ft/d)						Storage coefficient					
		Horizontal			Vertical			Specific storage (1/ft)			Specific yield (ft³/ft³)		
		Upper	Middle	Lower	Upper	Middle	Lower	Upper	Middle	Lower	Upper	Middle	Lower
Surprise Spring [1]	Outer	22	22	1	2.2	2.2	0.1	NA	NA	1.00E–06	0.16	0.16	0.05
	Middle	25	25	1	2.5	2.5	.1	NA	NA	1.00E–06	.16	.16	.05
	Inner	33	33	1	3.3	3.3	.1	NA	NA	1.00E–06	.25	.25	.05
Joshua Tree [2]		60	5	.5	.6	.05	.005	NA	1.00E–06	1.00E–06	.2	NA	NA

[1] Surprise Spring model by Londquist and Martin (1991).

[2] Joshua Tree model by Nishikawa and others (2004).

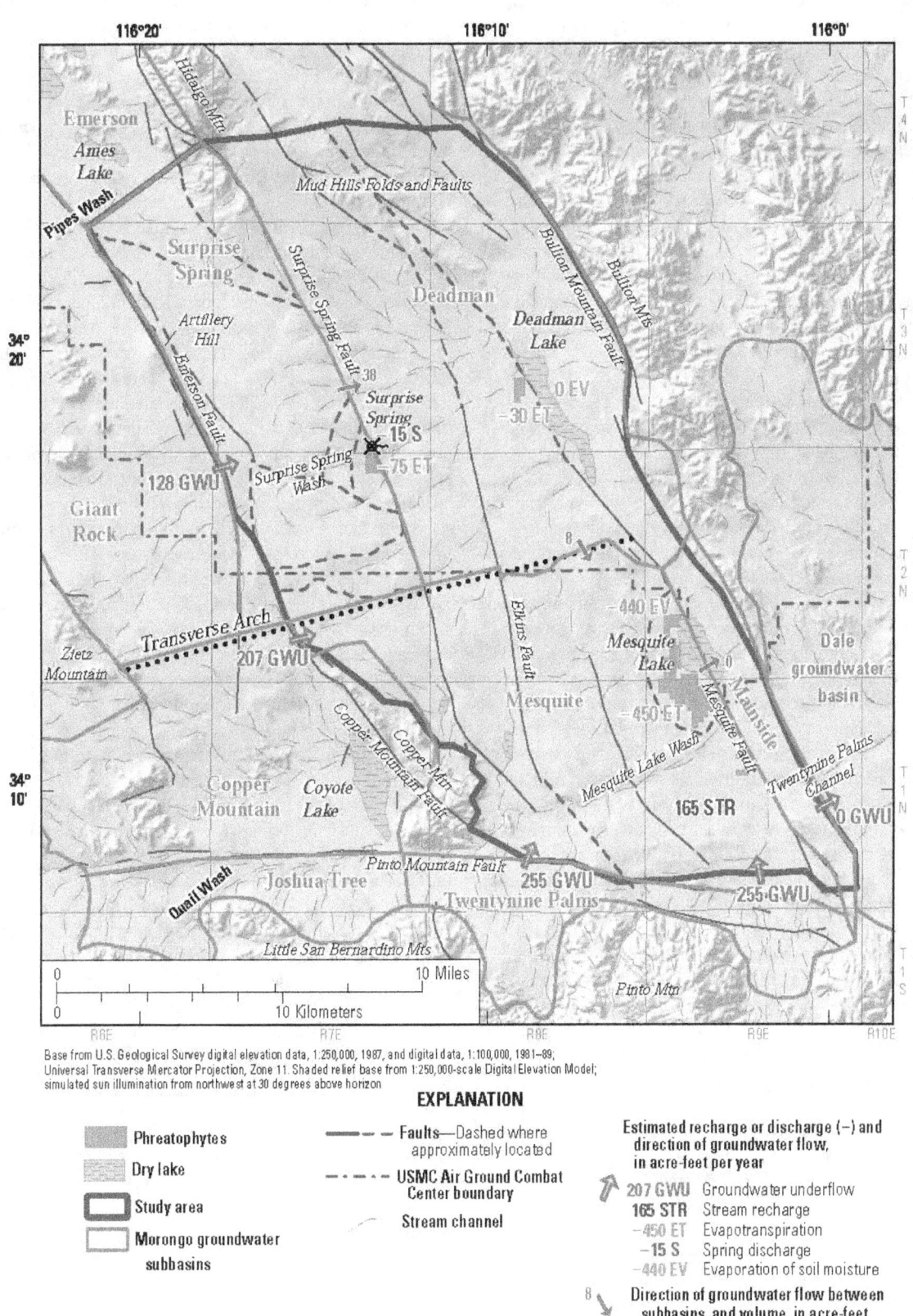

Figure 9.   Estimated water-budget components for steady-state conditions in the Twentynine Palms area, California.

precipitation over the basin is negligible because the average annual precipitation for the area only is about 4 to 6 in. and generally is not enough to meet evapotranspiration and soil-moisture requirements. Groundwater discharges from the study area naturally by spring flow, groundwater underflow to downstream basins, transpiration by phreatophytes, and direct evaporation from moist soil.

The Surprise Spring subbasin is recharged by groundwater underflow from the Giant Rock subbasin across the Emerson Fault that forms the western boundary of the Surprise Spring subbasin. (fig. 9). The source of the groundwater underflow is runoff from the San Bernardino Mountains that infiltrates the permeable deposits along Pipes Wash and its tributaries (fig. 2). Londquist and Martin (1991) estimated that the quantity of groundwater underflow was about 128 acre feet per year (acre-ft/yr) before groundwater pumping began in the subbasin (fig. 9).

Predevelopment discharge from the Surprise Spring subbasin consisted of flow from Surprise Spring, evapotranspiration by mostly mesquite trees near the spring, and groundwater underflow across the Surprise Spring Fault into the Deadman subbasin. Londquist and Martin (1991) estimated that the spring discharge was 15 acre-ft/yr, evapotranspiration was 75 acre-ft/yr, and groundwater underflow across the fault was 38 acre-ft/yr.

The Deadman subbasin is recharged by groundwater underflow from the Surprise Spring subbasin (fig. 9). Natural discharge from the subbasin includes evapotranspiration from mesquite trees along the west side of Deadman Lake (dry) and groundwater underflow into the Mesquite subbasin. Riley and Worts (1953 [2001]) estimated that steady-state (predevelopment) evapotranspiration in the area around Deadman Lake (dry) was about 30 acre-ft/yr and evaporation from the dry lakebed surface was negligible. Steady-state groundwater underflow to the Mesquite and the Mainside subbasins was estimated to be a total of 8 acre-ft/yr by balancing the total discharge from the subbasin with the estimated total recharge to the subbasin (recharge is equal to discharge under steady-state conditions).

The Mesquite subbasin is recharged primarily from groundwater underflow from the Copper Mountain subbasin to the west and the Twentynine Palms subbasin to the south (fig. 9). Runoff from the Little San Bernardino Mountains is the primary source of the recharge to the Copper Mountain subbasin, and ultimately to the Mesquite subbasin. Simulated predevelopment water budgets from a groundwater flow model of the Joshua Tree area by Nishikawa and others (2004) (fig. 7) suggest that approximately 207 acre-ft/yr of groundwater underflow discharged from the Copper Mountain subbasin into the Mesquite subbasin south of the Transverse Arch (figs. 7, 9).

Runoff from the Little San Bernardino and Pinto Mountains recharges the Mesquite subbasin as infiltration of streamflow that reaches the subbasin during storm events and as groundwater underflow resulting from the infiltration of streamflow in the Twentynine Palms subbasin. Net infiltration of runoff over the study area was estimated by using a regional watershed model developed for the Joshua Tree area (Nishikawa and others, 2004) (fig. 9). Net infiltration or direct recharge within the Surprise Spring, Deadman, Mesquite, and Mainside subbasins was simulated to be about 165 acre-ft/yr; most of the infiltration occurred along the Mesquite Lake Wash and Twentynine Palms Channel (Joseph Hevesi, U.S. Geological Survey, written commun., 2004). The quantity of underflow across the Pinto Mountain Fault from the Twentynine Palms subbasin is unknown.

Predevelopment discharge from the Mesquite subbasin consisted of evapotranspiration by predominately mesquite trees near the spring, evaporation from the Mesquite Lake (dry) moist playa surface, and possibly groundwater underflow through the Mesquite Fault into the Mainside subbasin (fig. 9). Riley and Worts (1953 [2001]) estimated that the transpiration loss in the Mesquite subbasin by about 2,000 acres of predominately mesquite trees near Mesquite Lake (dry) was approximately 450 acre-ft/yr. Evaporation from about 340 acres of seasonally moist soil near the Mesquite Lake (dry), containing saltgrass and mesquite trees, was estimated to be about 340 acre-ft/yr, and evaporation from the playa surface was estimated to be about 100 acre-ft/yr (Riley and Worts, 1953 [2001]). The total evapotranspiration in Mesquite subbasin was estimated to be about 890 acre-ft/yr before significant groundwater pumping began. In addition to discharge by evapotranspiration from the subbasin, there probably was a minor amount of discharge as groundwater underflow across the Mesquite Fault into the Mainside subbasin.

The quantity of groundwater underflow from the Twentynine Palms subbasin to the Mesquite subbasin was estimated to be 510 acre-ft/yr from a water-balance calculation for the Mesquite subbasin. Total discharge from the subbasin (890 acre-ft/yr) was about 510 acre-ft/yr greater than the sum of the estimated groundwater underflow from the Copper Mountain (207 acre-ft/yr) and Deadman (8 acre-ft/yr) subbasins and the infiltration of streamflow within the subbasin (165 acre-ft/yr). This deficiency in the water balance was assumed to be the quantity of groundwater underflow from the Twentynine Palms subbasin. The groundwater underflow was assumed to be distributed equally beneath the Mesquite Lake Wash and the Twentynine Palms Channel where they cross the Pinto Mountain Fault (fig. 9).

Total natural recharge to or discharge from the Surprise Spring, Deadman, Mesquite, and Mainside subbasins is estimated to be about 1,010 acre-feet/yr. About 90 percent of the recharge originates as runoff from Little San Bernardino and Pinto Mountains to the south, and 10 percent originates as runoff from the San Bernardino Mountains to the west. About 80 percent of the estimated groundwater discharge occurs as evapotranspiration in the area near Mesquite Lake (dry).

Groundwater Pumping and Irrigation-Return Flow

Groundwater pumping in the study area includes domestic pumping in the Mesquite subbasin, the MCAGCC pumping in the Surprise Spring, Deadman, and Mainside subbasins, and municipal pumping by the Twentynine Palms Water District (TPWD) in the Mesquite subbasin.

Before the 1950s, only a small amount of domestic groundwater was used in the Mesquite subbasin. A report compiled by Haley and Aldrich (2001) for TPWD indicated that about 350 private wells had been constructed in the Mesquite subbasin. Historical pumpage records for these private wells are not available; however, service connection water-delivery data suggests that production from individual private wells may average from 350 to 500 gallons per day per residential unit (Haley and Aldrich, 2001). Assuming half of the private wells constructed were active during any one year, the maximum private well pumpage for any given year between 1950 and 1999 can range from about 22 to 32 million gallons, or 70 to 100 acre-feet (acre-ft).

The MCAGCC drilled their first two supply wells (2N/7E –3A1 and 2N/7E–3B1) in the Surprise Spring subbasin near Surprise Spring in 1953 to provide water for the base (fig. 8). By 1970, another three wells (3N/7E–35P2, 2N/7E–2D1, and 2N/7E–3E1) were drilled within 2 mi of the Surprise Spring (fig. 8). In 1978, the MCAGCC constructed three more wells (3N/7E–28R1, 3N/7E–29R1, and 3N/7E–32J1) northwest of the original well field to mitigate groundwater declines near the spring; these wells were not put into operation until 1980 (Londquist and Martin, 1991). Three additional wells (3N/7E–28D1, 3N/7E–29F1, and 3N/7E–32D1) were drilled in 1991 north and west of existing supply wells. In 2000, a replacement well was drilled near well 2N/7E–3A1; this well was not used until 2002.

Pumpage from metered MCAGCC supply wells in the Surprise Spring subbasin has generally increased from 1953 to 2007 (fig. 10). The cumulative volume of groundwater pumped from these wells was about 139,400 acre-ft. Annual average pumpage from 1953 to 2007 was about 2,530 acre-ft. All of the water pumped from the Surprise Spring subbasin was transferred out of the subbasin for water supply in the Mainside subbasin.

Groundwater was pumped from the Deadman subbasin for local military uses for several years in the 1960s. Pumpage records are incomplete, but available records from well 3N/8E–29L1 indicate that pumpage was minimal (fig. 10).

To meet summer demand for irrigating the MCAGCC golf course, a supply well was constructed in the Mainside subbasin in the early 2000s. The pumpage record from this well is incomplete, but MCAGCC estimated that the well produced about 540 acre-ft in 2008 (Robert Lehman, Chief of Engineering, Marine Corps Air Ground Combat Center, written commun., 2008). Groundwater pumping by the city of Twentynine Palms in the Mesquite subbasin began in 2003 from well 1N/9E–21H1 (fig. 8). The average annual pumpage rate was approximately 850 acre-ft/yr during 2003–2007.

Return flows from the irrigation of lawns and fields at MCAGCC in the Mainside subbasin and the golf course in the Mesquite subbasin are a potential source of recharge to these subbasins. Available water-level data for the Mainside subbasin indicates that water levels have been essentially stable during the past 15 years (Stamos and others, 2006). This stability of water levels suggests that the return flows have not yet reached the water table. The depth to water in the Mainside subbasin is greater than 200 ft (fig. 4B), and several clay lenses are present in the thick unsaturated zone. The low permeability clay lenses probably reduce the vertical rate of movement of the irrigation return flow. Although irrigation return flows are not currently (2010) a significant source of recharge, this source of water will eventually reach the water table and could be a significant source of water for the Mainside subbasin in future years.

Water levels in well 2N/9E–19D2, near the MCAGCC golf course in the Mesquite subbasin, have risen about 15 ft from 2000 to 2007 (fig. 11). The increase in water levels in this well suggests that irrigation-return flows from the golf course are recharging the aquifer system. However, water levels in golf course well 2N/9E–19D4, have remained stable during this same period. The discrepancy in water-level change between the two wells may be the result of compartmentalization caused by fault segments of the Mesquite Fault. There was more than a 70 ft difference in water levels in these closely spaced wells prior to 2000 (fig. 11), suggesting that a barrier may separate the wells. Well 2N/9E–19D2 probably is in a small compartment formed by the Mesquite Fault to the east and an unnamed fault to the west; therefore, a small amount of recharge from the return flows could cause a rapid rise in water levels.

Groundwater Conditions

Since the MCAGCC was established in the 1950s, more than 10,000 water-level and water-quality measurements made by the USGS and other agencies have been stored in the USGS National Water Inventory System (NWIS) database. These data were used in this study to evaluate groundwater conditions in the Twentynine Palms area.

Groundwater Levels

Water-level contours representing 2006 conditions (Stamos and others, 2007) indicate that water-level altitudes measured in an individual subbasin are similar; however, water-level altitudes vary significantly across subbasin bounaries. Water-level altitudes descend in a stair-step manner from the uppergradient to the downgradient groundwater subbasins (fig.12). The highest water-level altitudes are near areas of recharge at the southern and western edges of the study area, and the lowest altitudes are near discharge areas at the eastern edge.

Sharp discontinuities in water-level altitudes on opposite sides of the Surprise Spring and Mesquite Faults and, to a

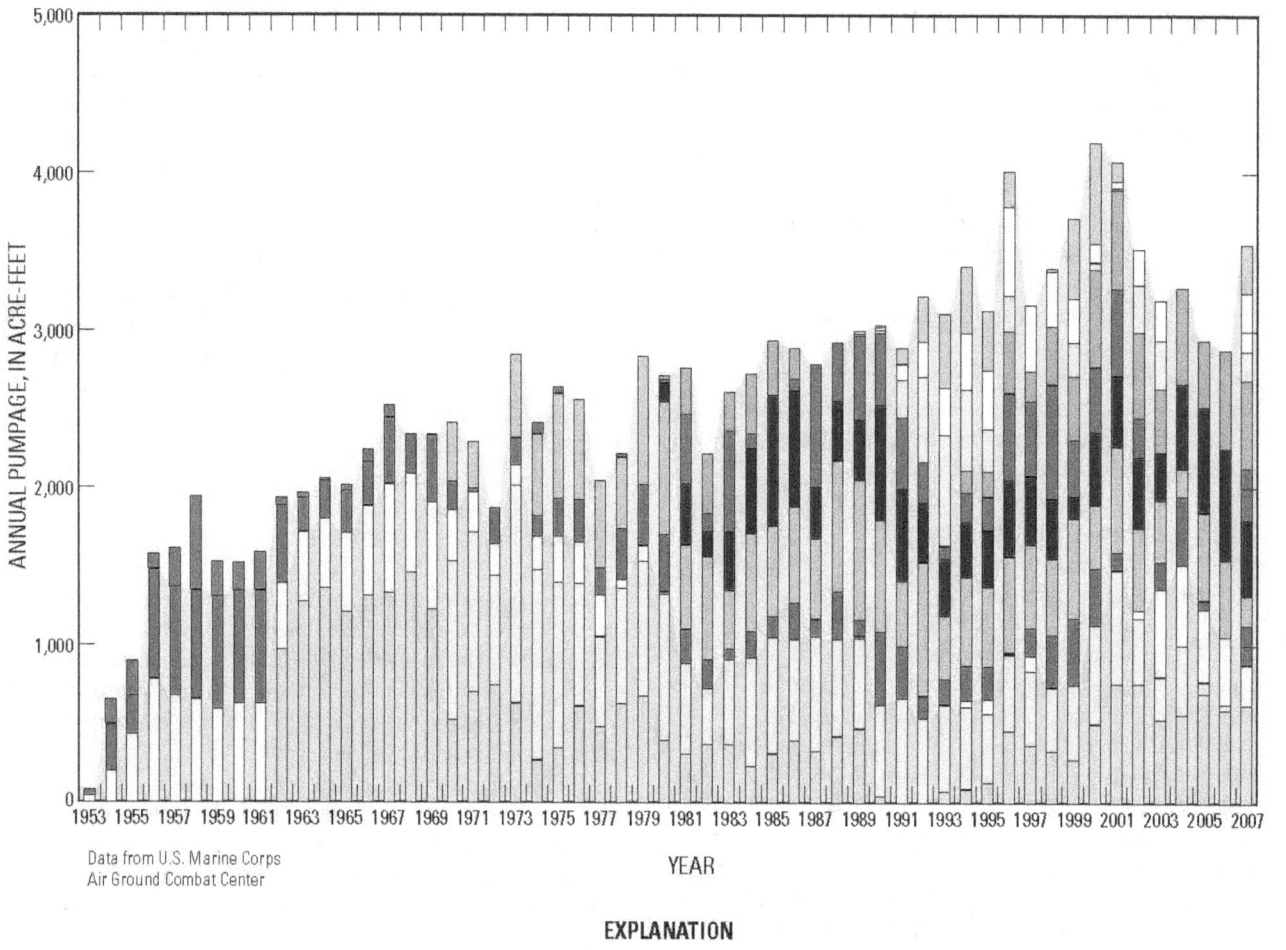

YEAR

EXPLANATION

Surprise Spring subbasin—

	State well name	Local name	Zone
	3N/7E-32D2	SW12A	
	3N/7E-29F1	SW11A	
	3N/7E-28D1	SW10A	Outer
	3N/7E-29R1	SW9A	
	3N/7E-32J1	SW8A	
	3N/7E-28R1	SW7A	
	2N/7E-3E1	SW6A	Middle
	2N/7E-3B1	SW2A	
	2N/7E-3A1	SW3A	
	2N/7E-2D1	SW5A	Inner
	3N/7E-35P2	SW4A	

Deadman subbasin—

	3N/8E-29L1	SW1A	West

Figure 10. Annual pumpage of U.S. Marine Corps Air Ground Combat Center (MCAGCC) supply wells in the Surprise spring subbasin, California, 1953–2007.

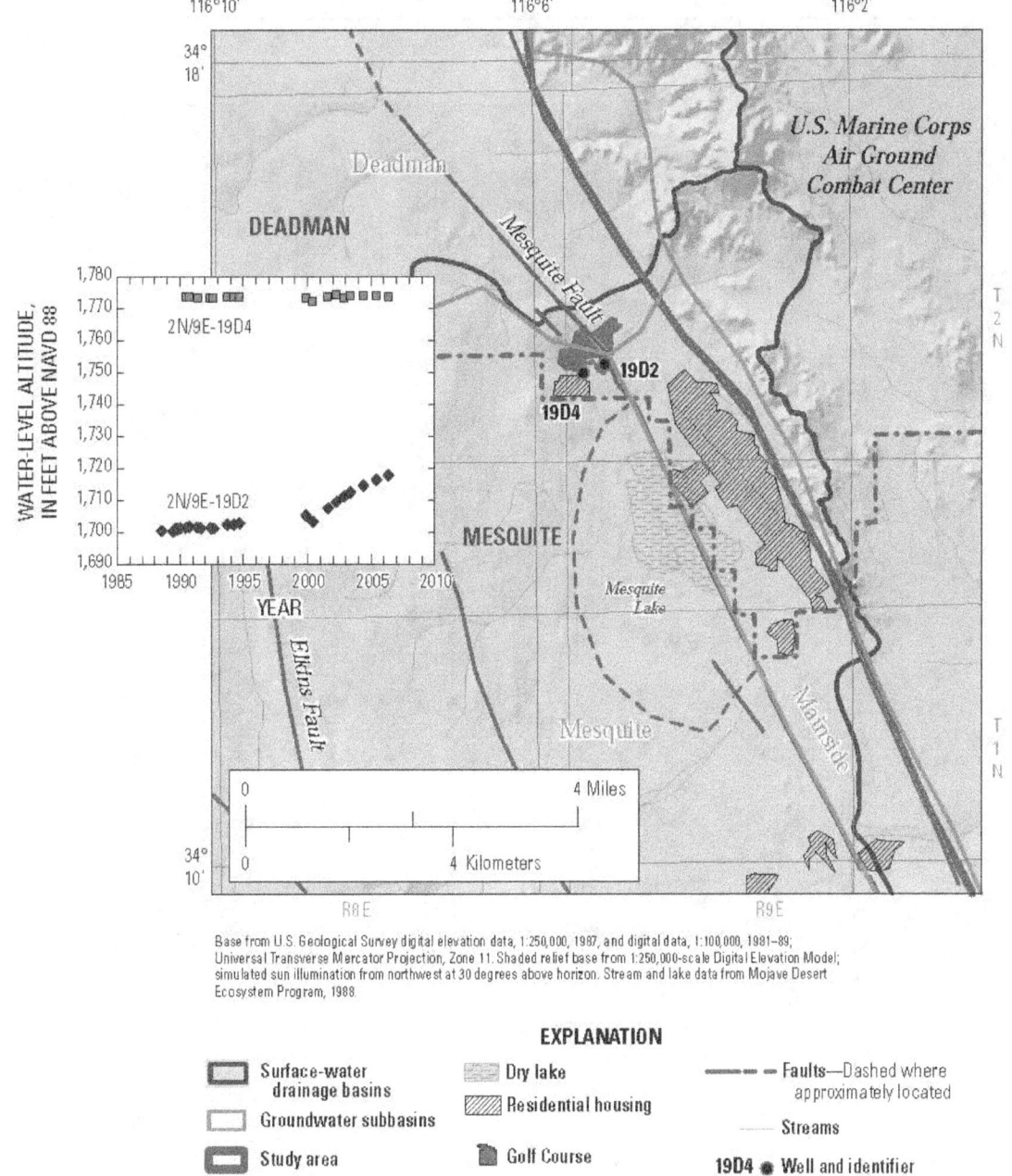

Base from U.S. Geological Survey digital elevation data, 1:250,000, 1987, and digital data, 1:100,000, 1981–89;
Universal Transverse Mercator Projection, Zone 11. Shaded relief base from 1:250,000-scale Digital Elevation Model;
simulated sun illumination from northwest at 30 degrees above horizon. Stream and lake data from Mojave Desert
Ecosystem Program, 1988.

EXPLANATION

Surface-water drainage basins	Dry lake	— ·· — Faults—Dashed where approximately located
Groundwater subbasins	Residential housing	—— Streams
Study area	Golf Course	**19D4** ● Well and identifier

Figure 11.   Location of the U.S. Marine Corps Air Ground Combat Center (MCAGCC) golf course and monitoring wells (Mesquite subbasin), with graph showing water-level altitudes at two nearby wells from 1988 to 2007, Twentynine Palms area, California.

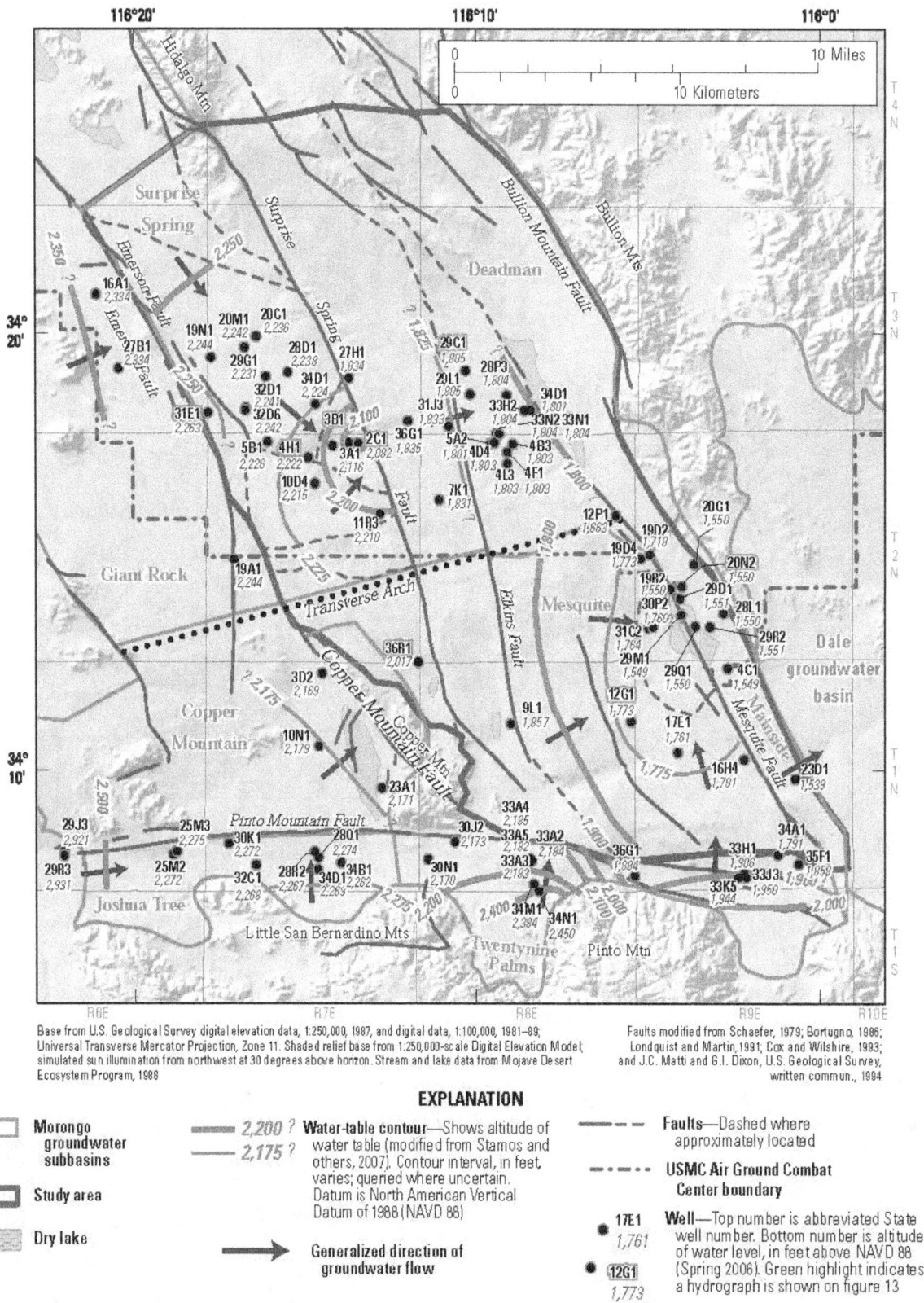

Base from U.S. Geological Survey digital elevation data, 1:250,000, 1987, and digital data, 1:100,000, 1981–89;
Universal Transverse Mercator Projection, Zone 11. Shaded relief base from 1:250,000-scale Digital Elevation Model;
simulated sun illumination from northwest at 30 degrees above horizon. Stream and lake data from Mojave Desert
Ecosystem Program, 1988

Faults modified from Schaefer, 1979; Bortugno, 1986;
Londquist and Martin, 1991; Cox and Wilshire, 1993;
and J.C. Matti and G.I. Dixon, U.S. Geological Survey,
written commun., 1994

EXPLANATION

	Morongo groundwater subbasins
	Study area
	Dry lake

2,200 ? **Water-table contour**—Shows altitude of
2,175 ? water table (modified from Stamos and
others, 2007). Contour interval, in feet,
varies; queried where uncertain.
Datum is North American Vertical
Datum of 1988 (NAVD 88)

→ Generalized direction of
groundwater flow

– – – **Faults**—Dashed where
approximately located

–·–·– **USMC Air Ground Combat
Center boundary**

17E1
1,761

12G1
1,773

Well—Top number is abbreviated State
well number. Bottom number is altitude
of water level, in feet above NAVD 88
(Spring 2006). Green highlight indicates
a hydrograph is shown on figure 13

Figure 12. Groundwater-level contours for 2006 for the Twentynine Palms area, California.

lesser extent, the Emerson and Elkins Faults indicate that these faults are barriers to groundwater flow (fig.12). As described in the "Geologic Structure" section of this report, the barrier effect of these faults is primarily caused by the low permeability of the fault zones resulting from the compaction and extreme deformation of the water-bearing deposits immediately adjacent to the faults, and by lateral juxtaposition of high- and low-permeability units. The low-permeability fault zones form a partial barrier to the lateral movement of groundwater flow, which can cause sharp discontinuities in water-level altitudes on opposite sides of a fault. In some cases, the barrier effect of the fault is not observed until the aquifer system is stressed by groundwater pumping or recharge.

Except for wells in the Surprise Spring subbasin, representative long-term hydrographs for wells in other part of the study area are nearly flat, suggesting that changes in groundwater storage are negligible (fig. 13). The long-term hydrographs for wells in the Surprise Spring subbasin show that water levels in wells near Surprise Spring declined almost immediately after groundwater development was initiated in the subbasin by the military in 1953 (figs. 10, 13A). The hydrograph of a well 2N/7E–2C1, adjacent to the Surprise Spring, shows the water level declining about 190 feet between 1953 and 2007, compared with about 60 ft at well 2N/7E–3B1. Different water-level altitudes and rates of water-level decline for wells in the Surprise Spring subbasin suggest the existence of a barrier to flow, such as a fault, between these wells (Londquist and Martin, 1991). The barrier effect of the faults prevents or reduces the lateral flow of water across the fault, which causes the water level to decline more on the side of fault where the water is being pumped than would be expected if the fault were not present.

Groundwater Quality

Water-quality maps of the historic highest reported measurements indicate that total dissolved solids, fluoride, and arsenic concentrations generally are lower in the Surprise Spring subbasin than in the Deadman, Mesquite, and Mainside subbasins (fig. 14). A similar map for chromium concentrations indicates that chromium concentrations are low (1–25 µg/L) except for a sample from one well in the southern part of the Surprise Spring subbasin (fig. 14D). Londquist and Martin (1991) reported that high concentrations of total dissolved solids, fluoride, and arsenic are indicative of water from the lower aquifer. Inspection of well depth and perforated intervals of the wells indicate that the high total dissolved solids, fluoride, and arsenic concentrations in the Surprise Spring subbasin are associated with wells in the lower aquifer. As discussed in the "Geohydrology" section of this report, the Surprise Spring subbasin is the only subbasin with a significant thickness of the upper aquifer. The water-quality data indicate that the upper aquifer contains water with lower total dissolved solids, fluoride, and arsenic concentrations than the middle and lower aquifers.

Groundwater-Flow Model

A regional-scale numerical groundwater-flow model was developed for the Surprise Spring, Deadman, Mesquite, and Mainside subbasins to better understand the aquifer system used by the MCAGCC for its water supply, and to provide a tool to help manage groundwater resources in the Twentynine Palms area. The model was used to test concepts about (1) the structure and hydraulic properties of the regional aquifer system, (2) the quantity and distribution of groundwater recharge and discharge in this area, (3) the hydraulic processes within and connections between the Surprise Spring, Deadman, Mesquite, and Mainside subbasins, and (4) the dynamics of groundwater flow in the four subbasins as an integrated system. After the model was determined to be a valid simulator of the regional groundwater-flow system, it was used to examine the potential effects of different water-supply and water-use strategies on groundwater conditions in the different subbasins and to evaluate the long-term water availability at a regional scale. In the future, the model also could be used to help identify the best water-management strategies for the MCAGCC by coupling it with optimization and particle tracking techniques.

The groundwater-flow model was developed by using MODFLOW–2000 (MF2K), a finite-difference computer code developed by the USGS (Harbaugh and others, 2000). A numerical groundwater-flow model is a set of equations that numerically describe groundwater status (in terms of hydraulic head and flow) in an aquifer system. A numerical model cannot, however, exactly duplicate the actual system because of uncertainties and the complex nature of the groundwater flow system and the limitations of numerical methods. Model development requires the use of assumptions and approximations that simplify the actual system. It cannot be overemphasized that the model only is as accurate as the assumptions and the data used in its development. To define the aquifer system numerically, the conceptual model of the aquifer system was divided into a lateral grid and vertical layers. Appropriate boundary conditions, hydraulic properties, and rate and distribution of recharge and discharge were estimated for the modeled aquifer system.

Model Discretization

Spatial Discretization

The active domain of the groundwater-flow model covers the entire Surprise Spring, Mesquite, and Mainside subbasins and the southern part of the Deadman subbasin (fig. 15). The aquifer system was simulated with three layers (fig. 16). The groundwater-flow model has a uniform grid consisting of 142 columns and 156 rows, with a total of 22,152 cells, each sized 820 ft by 820 ft, in each model layer (fig. 15). The grid of the regional model is parallel to the grid of the Universal

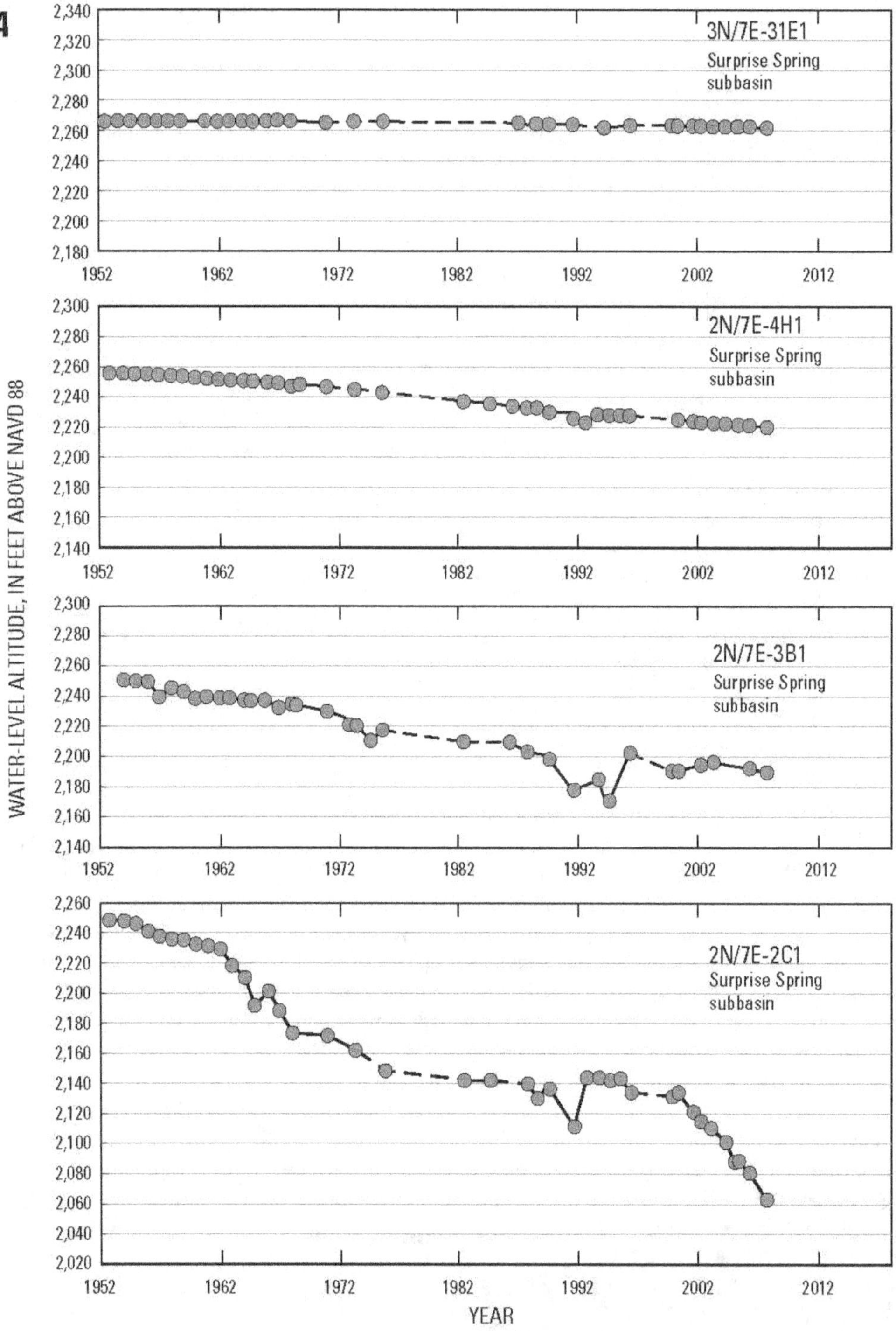

Hydrograph—Shows period of record for well. Dashed where data collection interval exceeds two years; dots indicate actual measurements

Figure 13. Water-level hydrographs for selected wells in (*A*) the Surprise Spring subbasin and (*B*) the Mesquite and Deadman subbasins of the Twentynine Palms area, California.

B

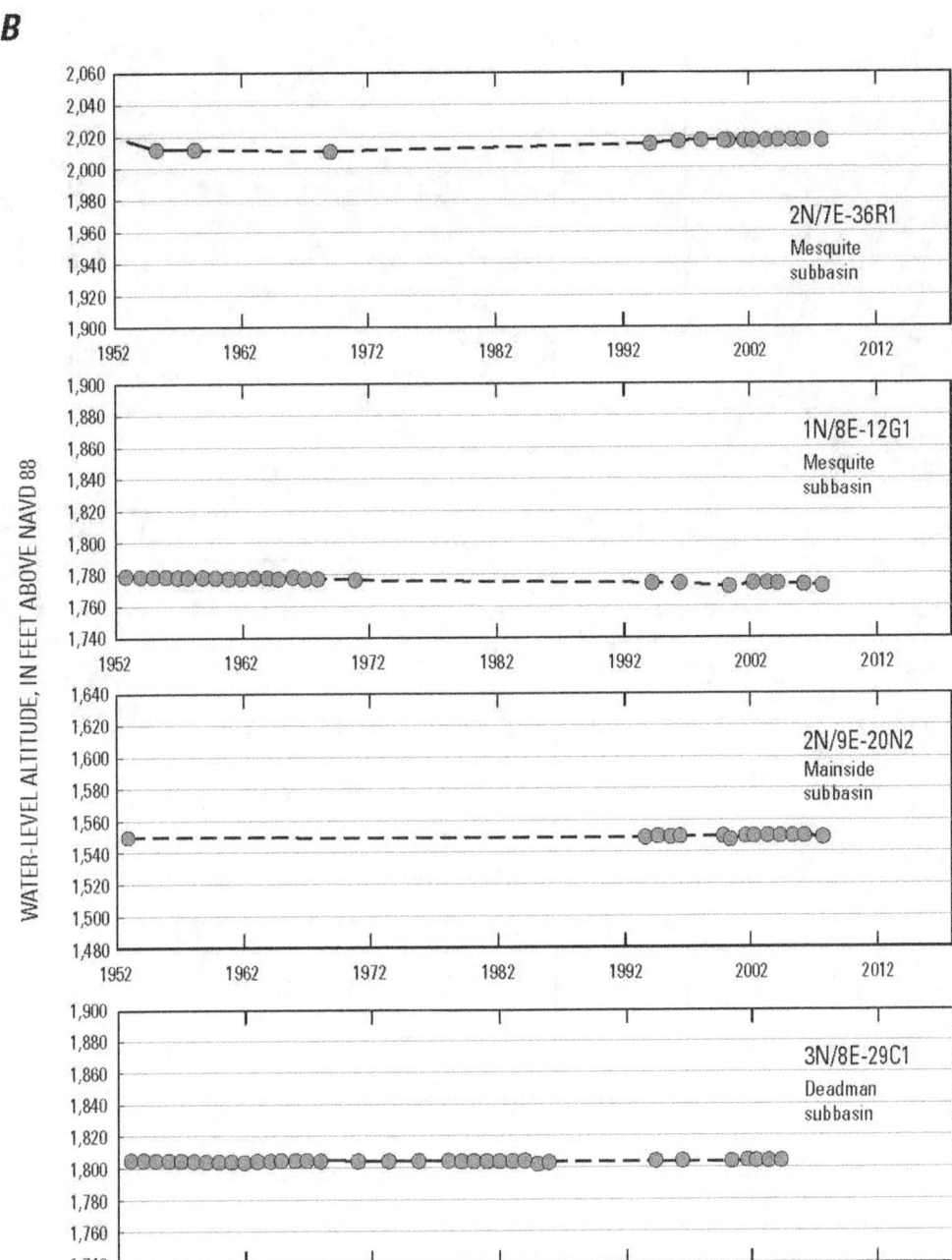

Hydrograph—Shows period of record for well. Dashed where data collection interval exceeds two years; dots indicate actual measurements

Figure 13. Continued.

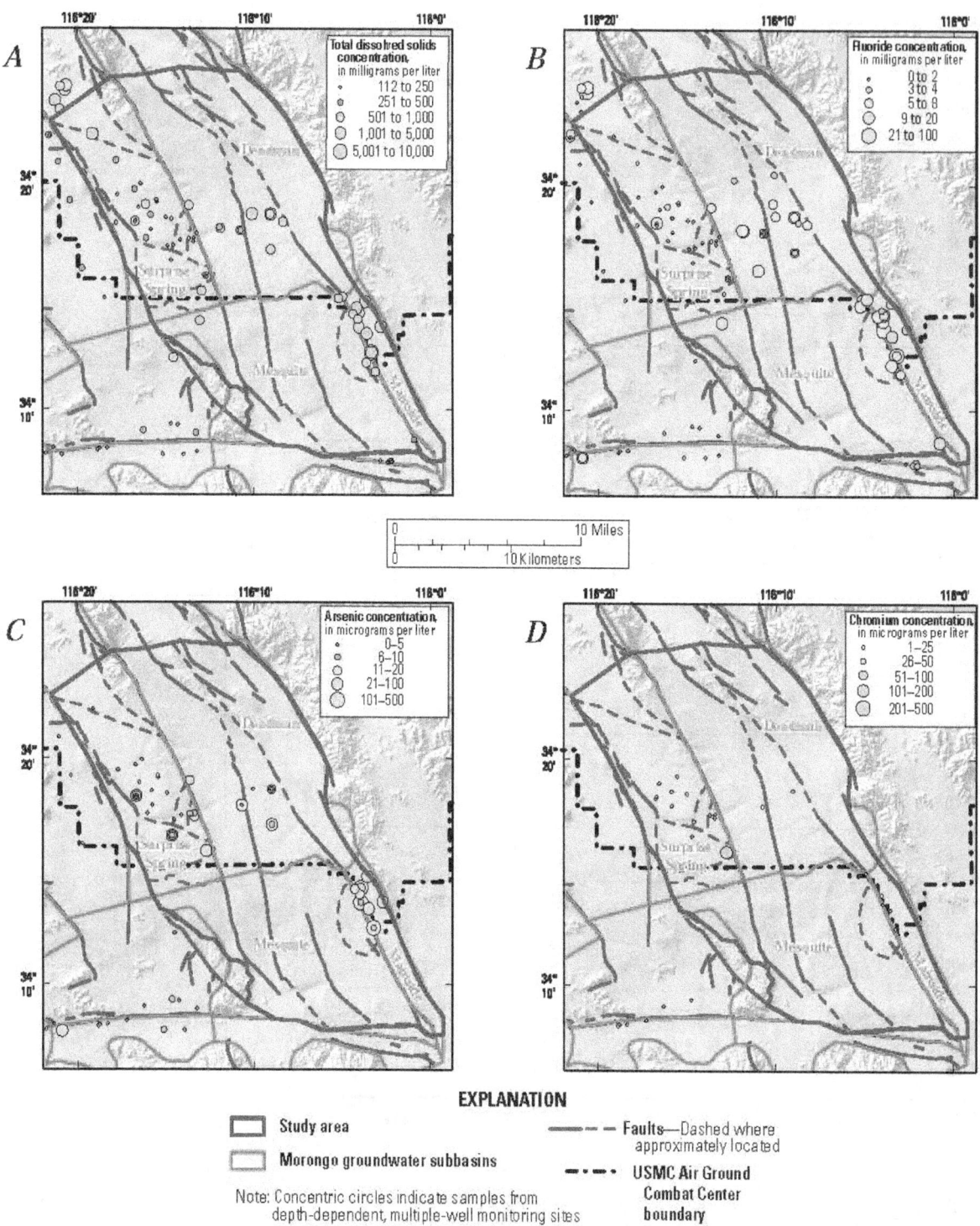

Figure 14.   Spatial distribution of historical highest (*A*) total dissolved solids, (*B*) fluoride, (*C*) arsenic, and (*D*) chromium concentrations in the Twentynine Palms area, California.

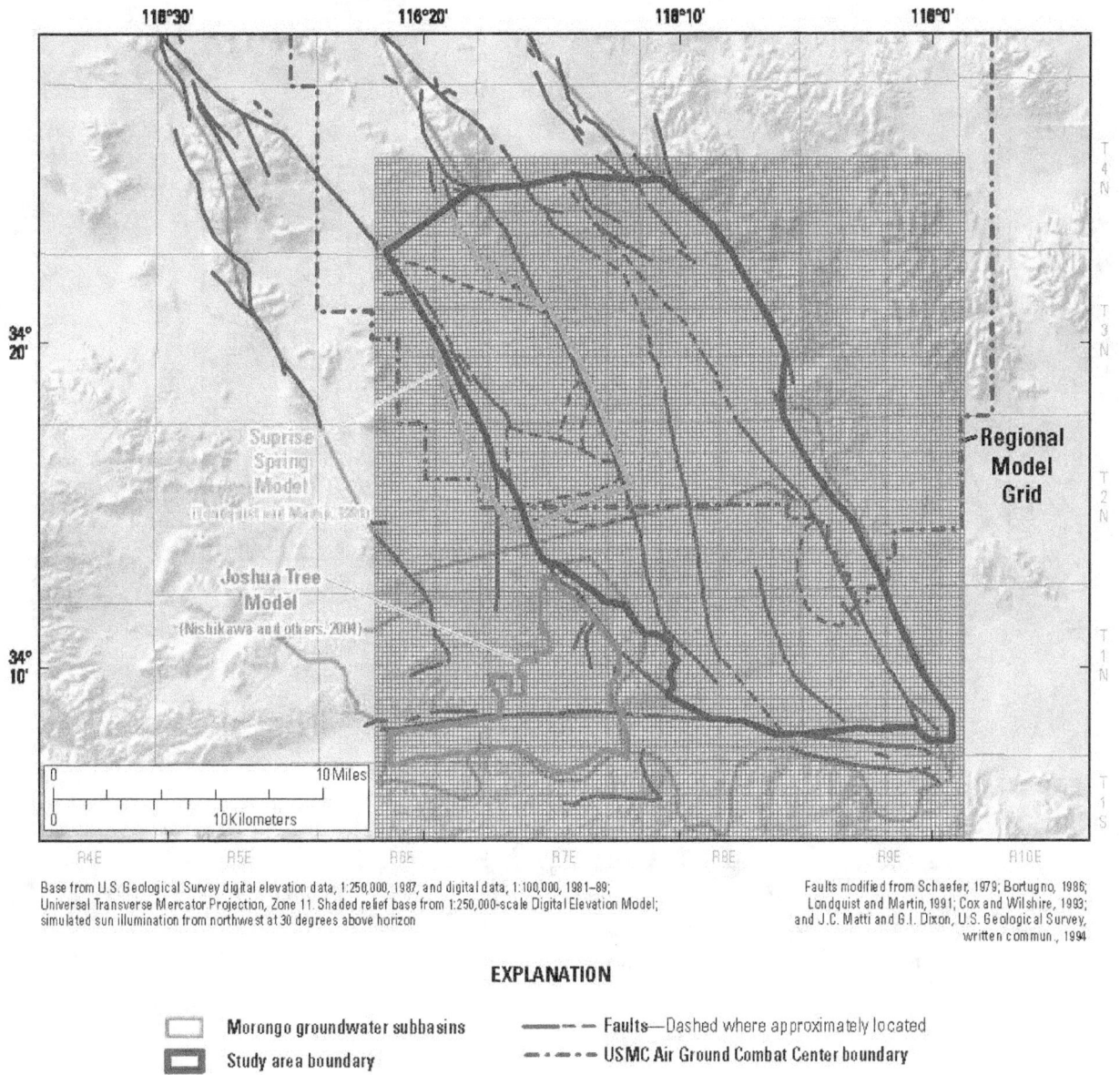

Base from U.S. Geological Survey digital elevation data, 1:250,000, 1987, and digital data, 1:100,000, 1981–89;
Universal Transverse Mercator Projection, Zone 11. Shaded relief base from 1:250,000-scale Digital Elevation Model;
simulated sun illumination from northwest at 30 degrees above horizon

Faults modified from Schaefer, 1979; Bortugno, 1986;
Londquist and Martin, 1991; Cox and Wilshire, 1993;
and J.C. Matti and G.I. Dixon, U.S. Geological Survey,
written commun., 1994

EXPLANATION

☐ Morongo groundwater subbasins

▭ Study area boundary

— - — Faults—Dashed where approximately located

— · — · — USMC Air Ground Combat Center boundary

Figure 15. Model grid for the regional groundwater-flow model of the Twentynine palms area, and the active model domain for the previously published groundwater-flow models for the Surprise Spring subbasin and the Joshua Tree area, California.

A

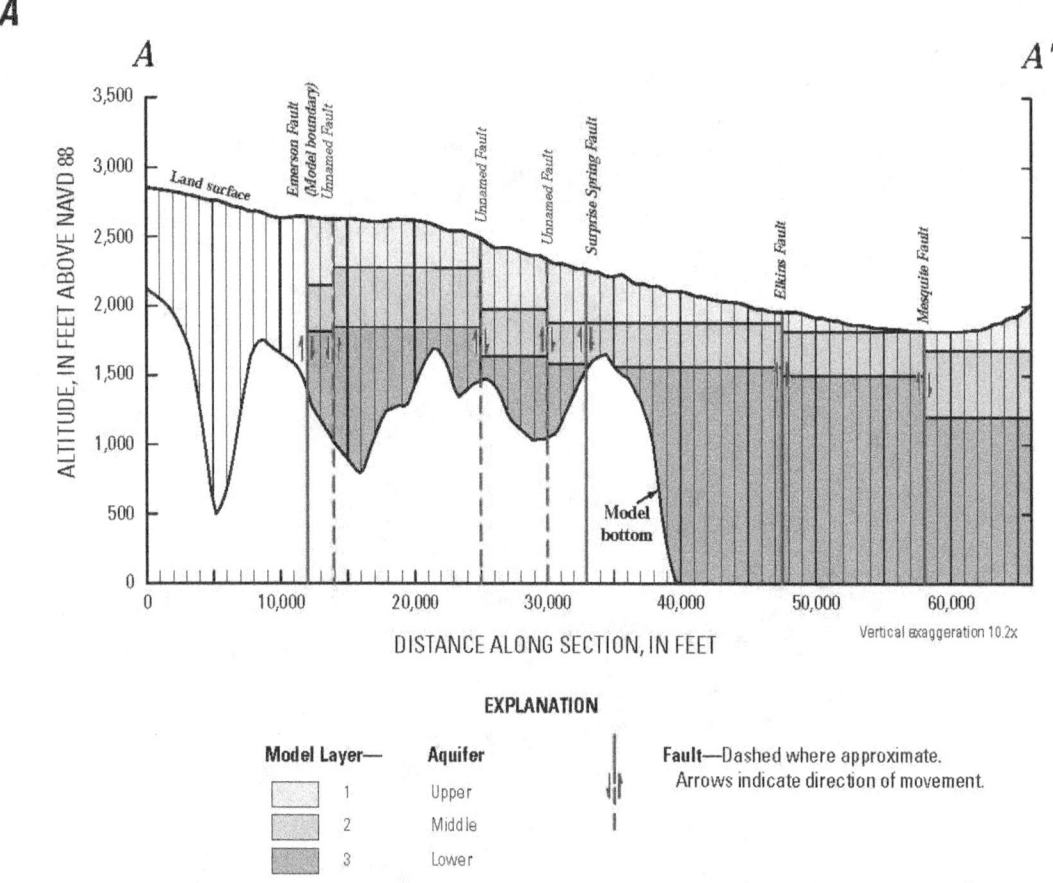

EXPLANATION

See figure 3 for location and figure 4A for corresponding geologic section.

Figure 16.   Generalized model cross section A–A', B–B', C–C and D–D' showing vertical discretization of the regional groundwater-flow model of the Twentynine Palms area, California.

B

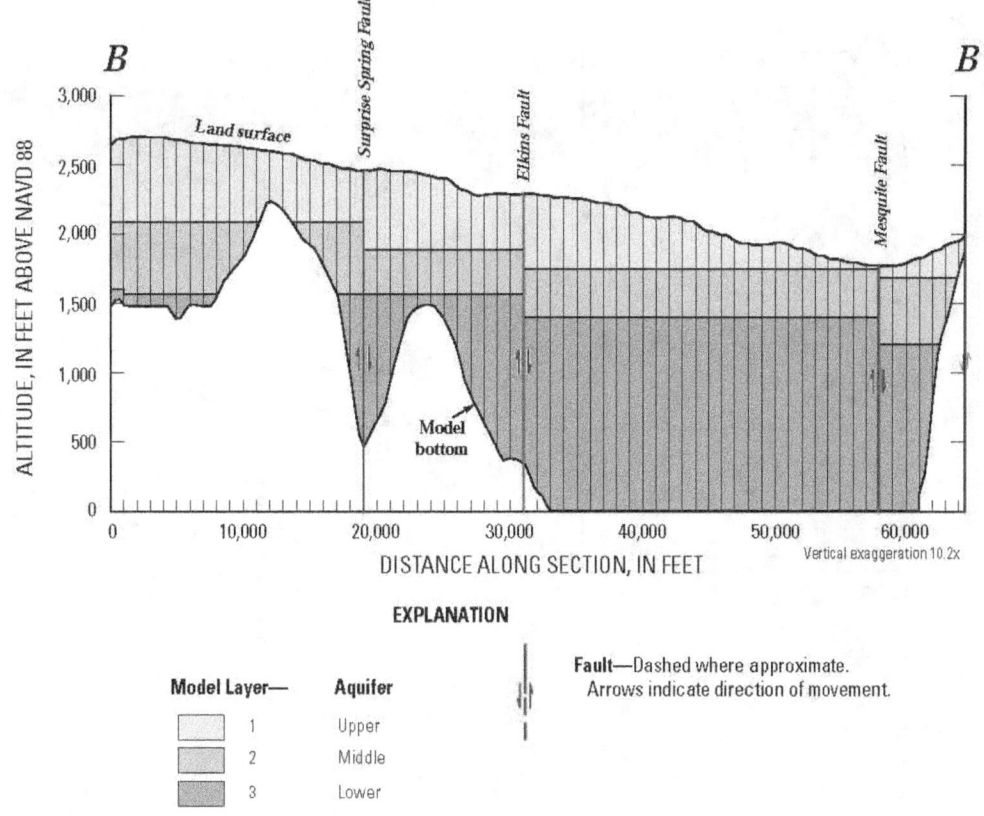

EXPLANATION

Model Layer—   Aquifer

1   Upper
2   Middle
3   Lower

Fault—Dashed where approximate.
Arrows indicate direction of movement.

See figure 3 for location and figure 4*B* for corresponding geologic section.

Figure 16. Continued.

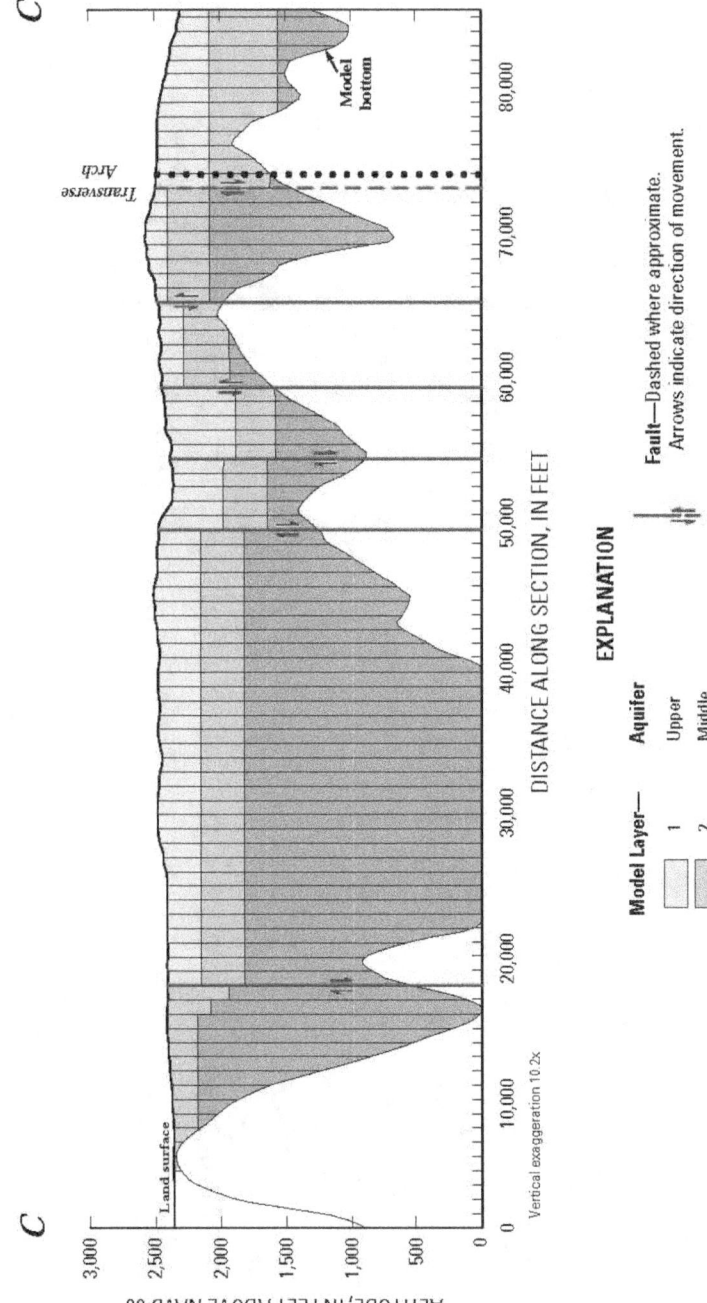

Figure 16.   Continued.

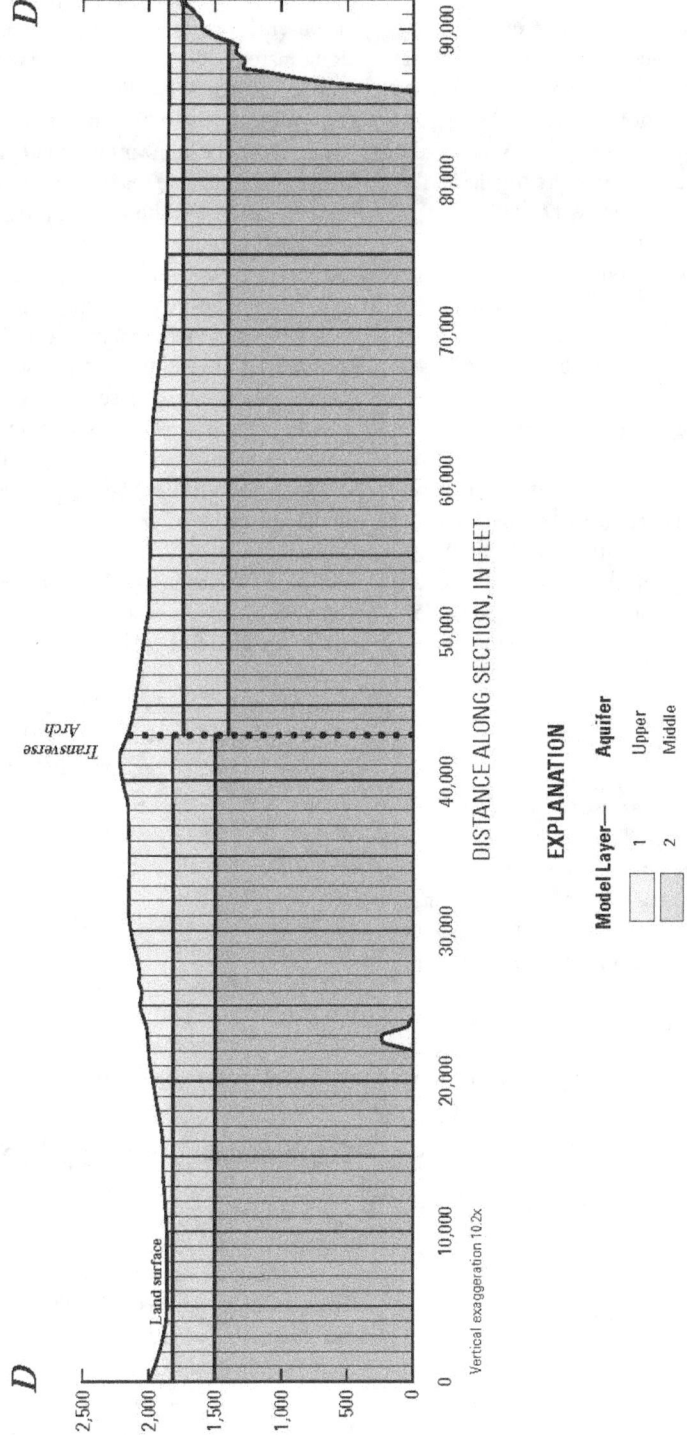

D
D'

Transverse
Arch

Land surface

ALTITUDE, IN FEET ABOVE NAVD 88

2,500
2,000
1,500
1,000
500
0

Vertical exaggeration 10.2x

0   10,000   20,000   30,000   40,000   50,000   60,000   70,000   80,000   90,000

DISTANCE ALONG SECTION, IN FEET

EXPLANATION

Model Layer—   Aquifer

1   Upper
2   Middle
3   Lower

See figure 3 for location and figure 4D for corresponding geologic section.

Figure 16.   Continued.

Transverse Mercator (UTM) coordinate system; the columns are orientated north-south and the rows east-west.

Model-layer 1 represents the upper aquifer, model-layer 2 the middle aquifer, and model-layer 3 the lower aquifer. The top altitude of the model was set equal to land surface altitude derived from a USGS Digital Elevation Model (DEM). The bottom altitude of model-layer 3 was set equal to 0 ft or to the altitude of basement complex (derived from gravity data) if the altitude of the top surface of the basement complex was greater than 0 ft above NAVD 88 (fig. 16).

The altitudes of the bottoms of model-layers 1 and 2 were assigned the approximate altitudes of the base of the upper and middle aquifers, respectively (figs. 4, 16). The bottoms of model-layers 1 and 2 are assumed to be flat between identified faults or geologic barriers. In areas where the altitude of the basement complex was higher than that of the bottom of model-layer 1 or 2, the altitude of the basement complex was used as the bottom of the respective model layer (fig. 16). If the altitude of the basement complex was higher than that of the top of the model layer, the respective model cell was inactive in that layer. During transient-model simulations, the water-table may rise to land surface and (or) drop into model-layer 3. As a result, the simulated saturated thickness of all model layers can vary over time.

Temporal Discretization

The groundwater-flow model was used to simulate steady-state and transient conditions. The steady-state condition represented the condition before 1953, which was assumed to be the predevelopment condition in the study area. Simulated steady-state hydraulic heads were used as the initial hydraulic heads for the transient simulation that represented conditions from 1953 to 2007.

The temporal discretization for the transient simulations consisted of 31 annual stress periods for 1953 through 1983, and 288 monthly stress periods for 1984 through 2007. A stress period is a time interval during which all external stresses, including recharge and pumping rate, are constant. The length of a stress period was determined on the basis of available pumpage data. The MCAGCC recorded pumpage annually before 1984 and monthly afterwards.

Boundary Conditions

Three types of boundary conditions were used in the regional groundwater-flow model: no-flow, specified-flux, and general-head conditions.

A no-flow boundary indicates that there is no exchange of water between the boundary cell and the area outside of the active model grid. All model cells along the boundary of the active model domain and at the base of the active model domain were simulated as no-flow boundaries, except for cells where specified-flux or general-head boundaries were

simulated (fig. 17). Most of these no-flow boundaries correspond to locations where interpreted gravity data indicate that the basement complex is at or near the water-table altitude or along major fault zones that were considered to be barriers to groundwater flow. No-flow boundary conditions were specified for the northern boundary of the model domain in the Deadman subbasin near the Mud Hills (fig. 17), even though gravity data indicate that the basement complex is extremely deep in this part of the subbasin (fig. 5). Folds and faults associated with the Mud Hills are considered to be complete barriers to groundwater flow in this part of the subbasin.

A specified-flux boundary indicates that water flows into or out of the model domain at a specified rate that remains constant for the entire stress period. Specified-flux boundary conditions were used in selected model cells in model-layer 2 to simulate groundwater underflow along stream channels that cross (1) the Emerson Fault on the western boundary of the Surprise Spring subbasin, (2) the Copper Mountain Fault on the northwestern boundary of the Mesquite subbasin, and (3) the Pinto Mountain Fault on the southern boundary of the Mesquite subbasin (figs. 17, 18). Specified-flux boundary conditions also were used to simulate natural stream recharge within the model domain. The distribution and quantity of the groundwater underflow and natural stream recharge is discussed in the "Simulated Recharge" section of the report.

A general-head boundary (GHB) simulates flow across the boundary at a rate proportional to the difference between the hydraulic head at the boundary and that assigned at a source outside of the boundary (McDonald and Harbaugh, 1988). The constant of proportionality is the hydraulic conductance, which can be specified or estimated using model calibration. GHB cells were located at the southeastern boundary of the Mainside subbasin to simulate groundwater underflow between the Mainside subbasin and the Dale groundwater basin to the east (fig. 17). Similar to specified-flux cells, GHB cells were placed in model-layer 2 because model-layer 1 is dry in that area of the model domain.

Simulated Aquifer Properties

Aquifer properties assigned to model cells include horizontal and vertical hydraulic conductivity, specific yield, specific storage, and hydraulic characteristics of flow barriers. The aquifer properties affect the rate at which groundwater moves through an aquifer, the volume of water in storage, and the rate and areal extent of groundwater level declines caused by pumping. Initial values of these properties were estimated from aquifer tests, geologic interpretation, and values used in published groundwater models of the Surprise Spring subbasin (Londquist and Martin, 1991) and the Joshua Tree and Copper Mountain subbasins (Nishikawa and others, 2004). The final values of these aquifer properties were determined during the model-calibration process using a trial-and-error approach under steady-state (predevelopment) and transient conditions.

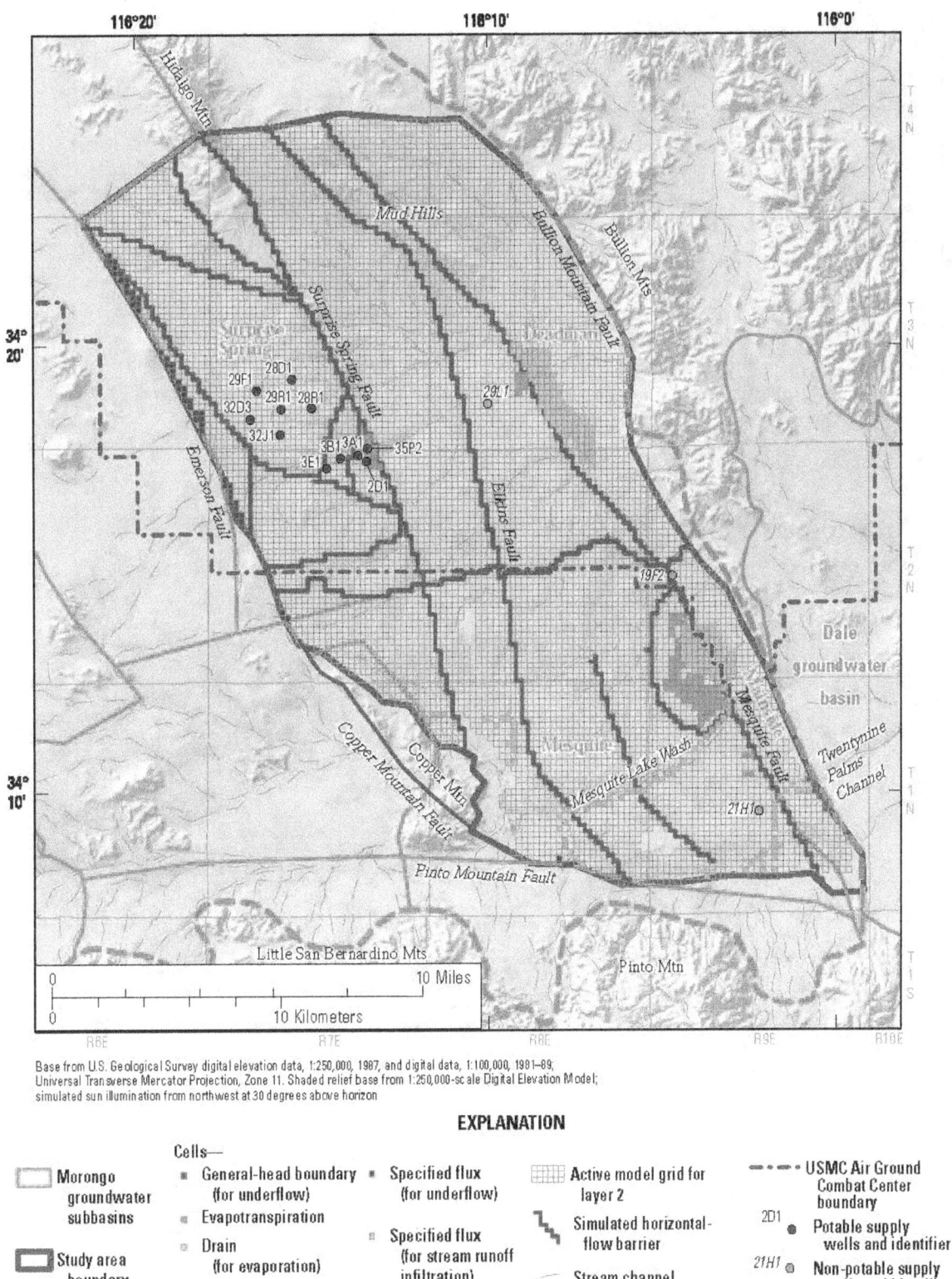

Base from U.S. Geological Survey digital elevation data, 1:250,000, 1987, and digital data, 1:100,000, 1981–89;
Universal Transverse Mercator Projection, Zone 11. Shaded relief base from 1:250,000-scale Digital Elevation Model;
simulated sun illumination from northwest at 30 degrees above horizon

EXPLANATION

Cells—
- Morongo groundwater subbasins
- Study area boundary
- General-head boundary (for underflow)
- Evapotranspiration
- Drain (for evaporation)
- Specified flux (for underflow)
- Specified flux (for stream runoff infiltration)
- Active model grid for layer 2
- Simulated horizontal-flow barrier
- Stream channel
- USMC Air Ground Combat Center boundary
- 2D1 Potable supply wells and identifier
- 21H1 Non-potable supply wells and identifier

Figure 17. Active model cells, general-head boundary cells, specified-flux boundary cells, evapotranspiration cells, drain cells, and supply wells simulated in the regional groundwater-flow model of the Twentynine Palms area, California.

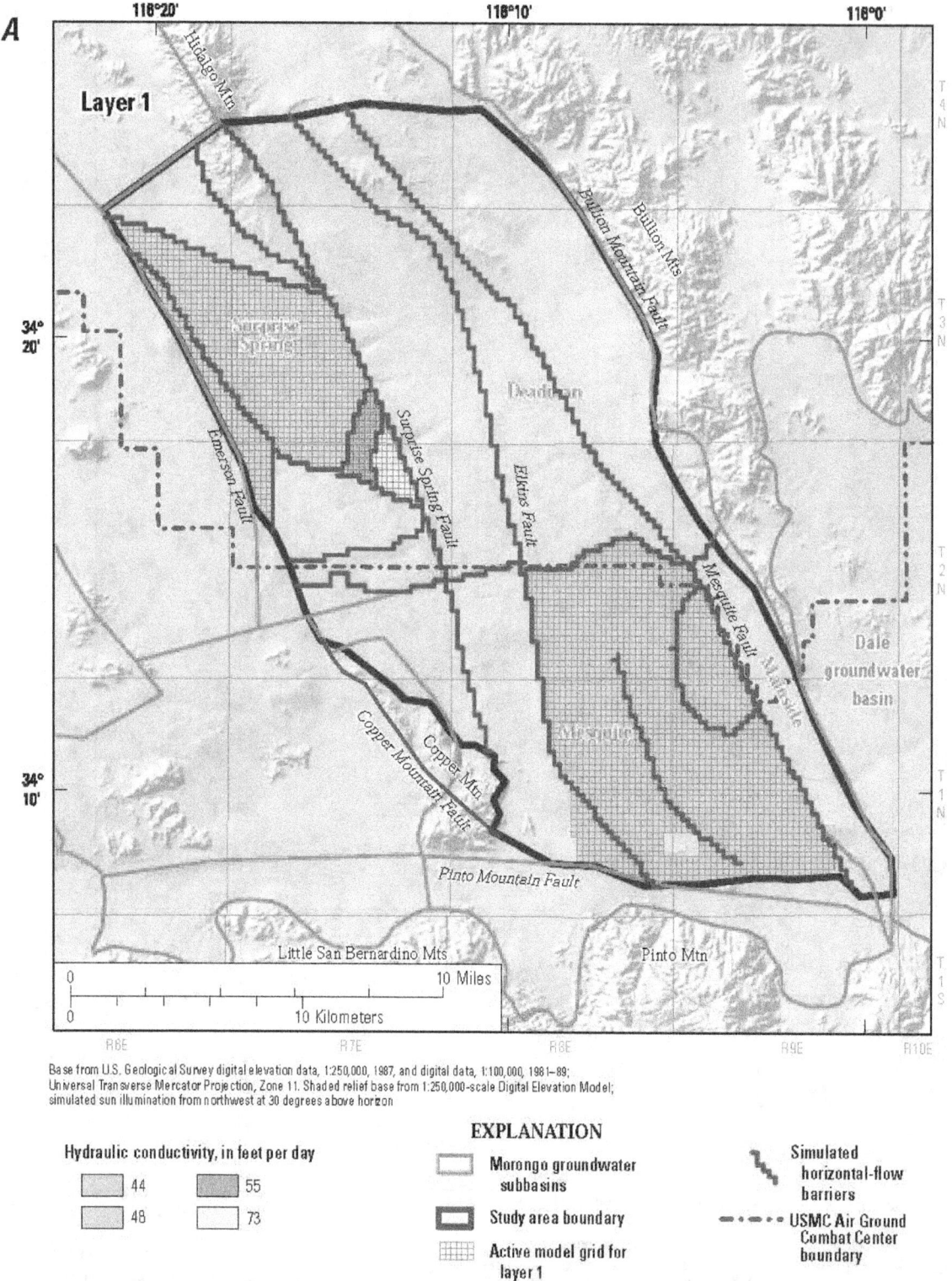

Base from U.S. Geological Survey digital elevation data, 1:250,000, 1987, and digital data, 1:100,000, 1981–89;
Universal Transverse Mercator Projection, Zone 11. Shaded relief base from 1:250,000-scale Digital Elevation Model;
simulated sun illumination from northwest at 30 degrees above horizon

EXPLANATION

Hydraulic conductivity, in feet per day

44		55	
48		73	

☐ Morongo groundwater subbasins

☐ Study area boundary

▦ Active model grid for layer 1

Simulated horizontal-flow barriers

–·–·– USMC Air Ground Combat Center boundary

Figure 18.   Distribution of horizontal hydraulic conductivity for (A) model layer 1, (B) model layer 2, and (C) model layer 3 of the regional groundwater-flow model of the Twentynine Palms area, California.

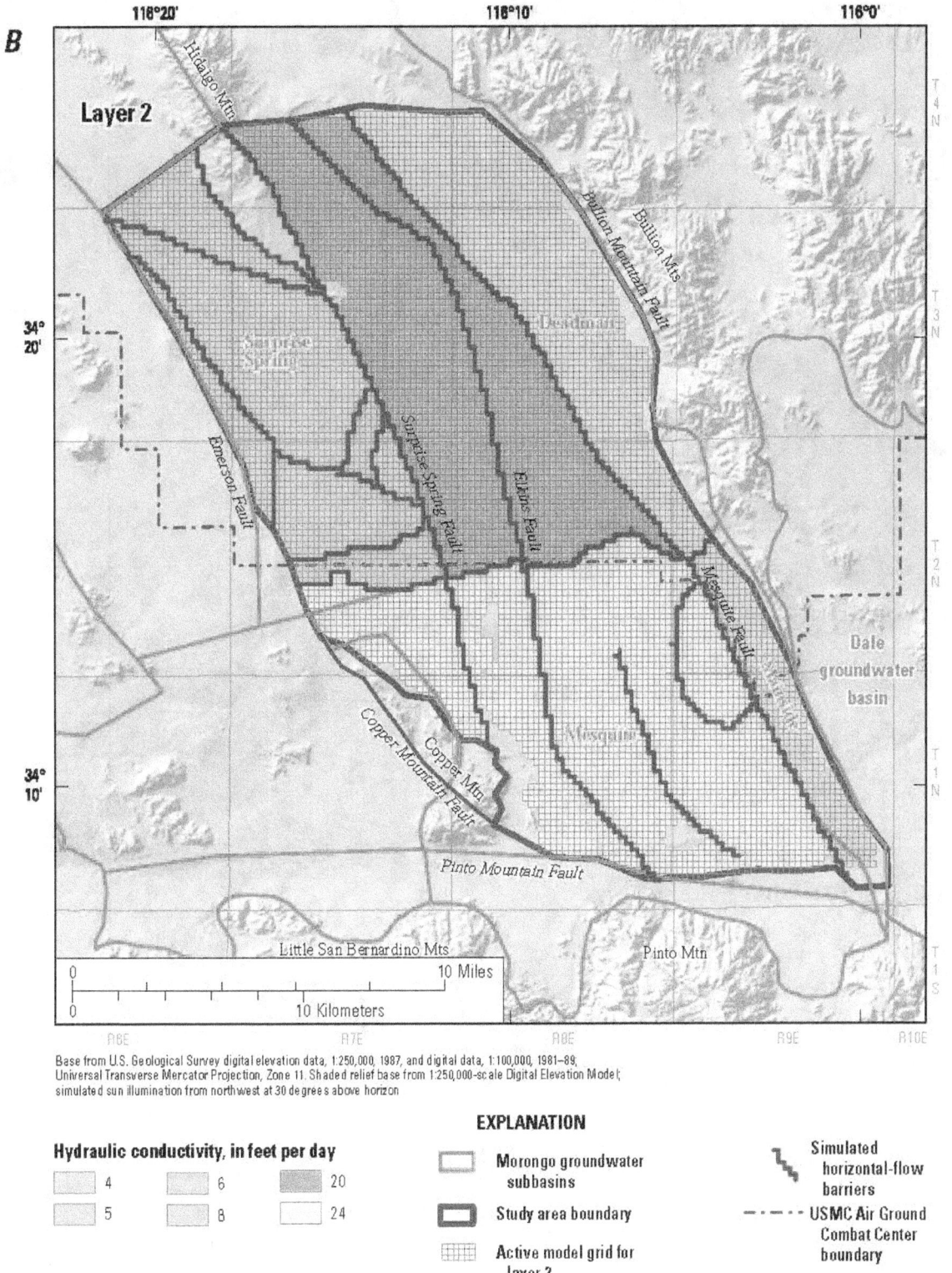

Base from U.S. Geological Survey digital elevation data, 1:250,000, 1987, and digital data, 1:100,000, 1981–89,
Universal Transverse Mercator Projection, Zone 11. Shaded relief base from 1:250,000-scale Digital Elevation Model;
simulated sun illumination from northwest at 30 degrees above horizon

EXPLANATION

Hydraulic conductivity, in feet per day

4	6	20
5	8	24

Morongo groundwater
subbasins

Study area boundary

Active model grid for
layer 2

Simulated
horizontal-flow
barriers

– · · – · · – USMC Air Ground
Combat Center
boundary

Figure 18.   Continued.

C

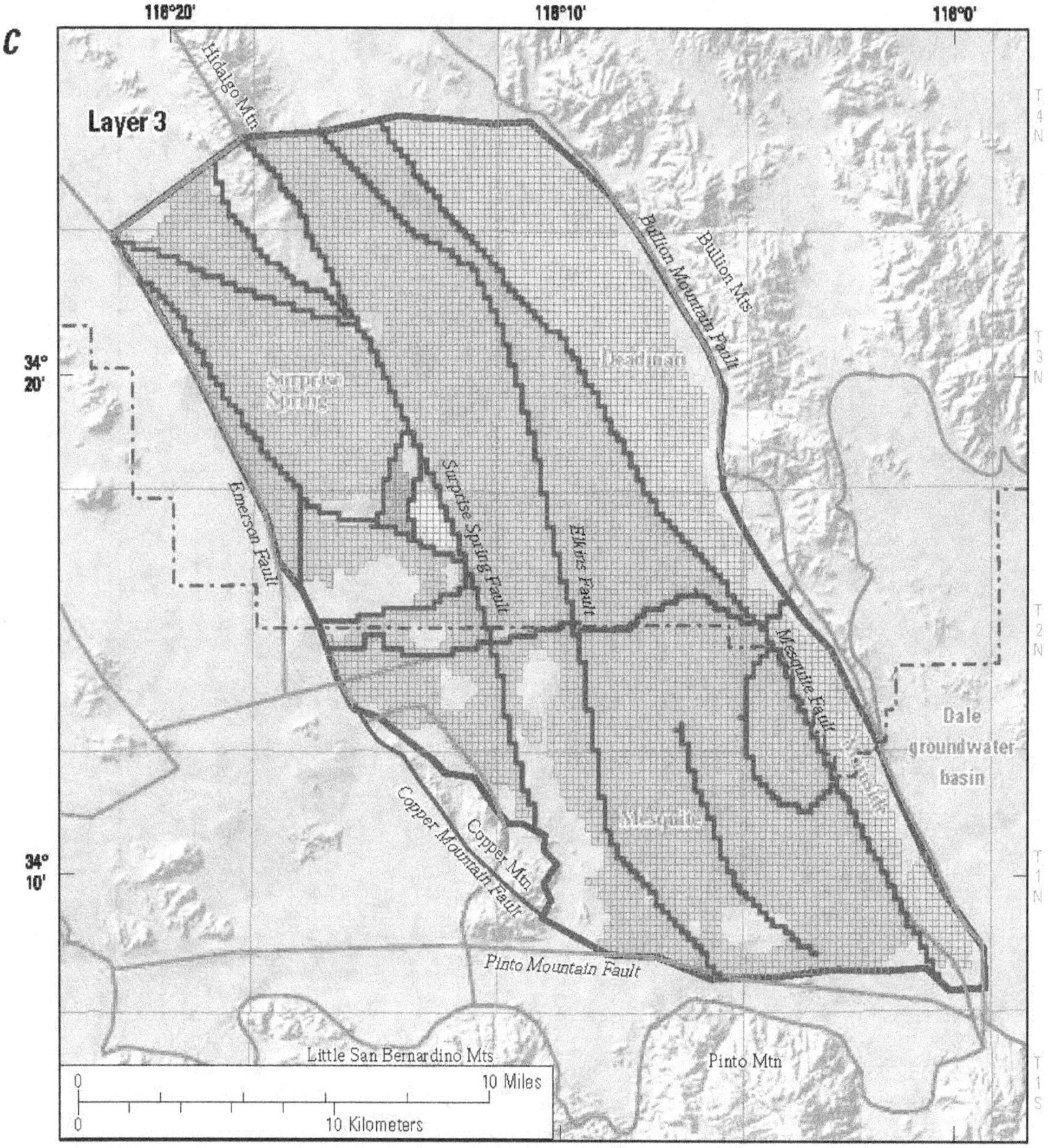

Layer 3

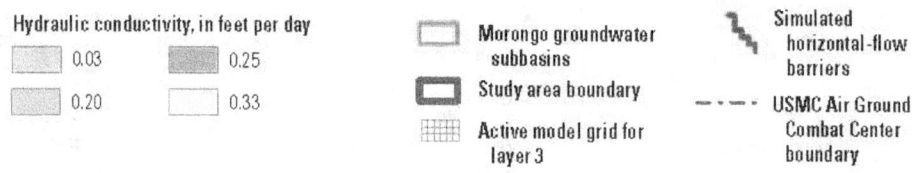

Base from U.S. Geological Survey digital elevation data, 1:250,000, 1987, and digital data, 1:100,000, 1981–89;
Universal Transverse Mercator Projection, Zone 11. Shaded relief base from 1:250,000-scale Digital Elevation Model;
simulated sun illumination from northwest at 30 degrees above horizon

EXPLANATION

Hydraulic conductivity, in feet per day

0.03   0.25

0.20   0.33

Morongo groundwater subbasins

Study area boundary

Active model grid for layer 3

Simulated horizontal-flow barriers

USMC Air Ground Combat Center boundary

Figure 18.   Continued.

Simulated Hydraulic Conductivity and Transmissivity

The initial estimates of hydraulic conductivity were derived from the specific-capacity values of local wells (table 1) and from calibrated models developed by the USGS for the Surprise Spring subbasin (Londquist and Martin, 1991) and the Joshua Tree area (Nishikawa and others, 2004) (fig. 15). Subsequent parameterization of the regional model was designed to reflect regional depositional patterns and local variations of basin sediments reported in previous studies. Each layer was divided into hydrogeologic zones with uniform hydraulic properties (fig. 7). The anisotropic ratio of horizontal and vertical hydraulic conductivity of each zone was assumed to be 100 to 1. The final calibrated hydraulic conductivity values are summarized in table 3, and the areal distribution of hydraulic conductivity for each layer is shown on figure 18.

The transmissivity of each model layer is the product of the horizontal hydraulic conductivity and the saturated thickness for each model cell. The calibrated transmissivity values for layers 1 and 2 are summarized in table 4. The values represent the transmissivity of each hydrogeologic zone at steady-state conditions.

Simulated Storage Properties

Calibrated specific-yield values were 0.25 in model-layer 1, 0.16 in model-layer 2, and 0.05 in model-layer 3 (table 3). Note, the specific yield of model-layers 2 and 3 only is used in the simulation if the overlying layer is unsaturated (dry). Calibrated specific-storage values for all zones were 3.0×10^{-6} in model-layers 1 and 2 and 1.0×10^{-6} in model-layer 3.

Simulated Fault Conductances

In the study area, faults are partial or complete barriers to groundwater flow. The Horizontal Flow Barrier (HFB) package (Hsieh and Freckleton, 1993) was used to simulate faults within the active model domain that impede the horizontal flow of groundwater (fig. 19). Faults are approximated as a series of HFBs conceptually situated between pairs of adjacent cells in the finite-difference grid (Hsieh and Freckleton, 1993). Flow across a simulated fault is proportional to the hydraulic-head difference between adjacent cells. The constant of proportionality is the hydraulic characteristic and is equal to the barrier hydraulic conductivity divided by the width of the horizontal-flow barrier. Initially, the hydraulic characteristic was set as a large value to allow groundwater to flow freely across the fault segments. During calibration, the hydraulic-characteristic values were lowered, as needed, such that simulated hydraulic heads closely matched measured water levels on both sides of the fault segment. The calibrated hydraulic-characteristic values of the HFB segments range from a high of 3.28×10^{-1} to a low of 3.28×10^{-11} per day (table 5).

Simulated Recharge

Natural groundwater recharge to the model domain occurs as underflow from adjacent groundwater subbasins and as infiltration of runoff along washes within the study area. Artificial recharge from irrigation-return flows is a possible source of recharge to the Mainside and Mesquite subbasins. Interpretation of water-level data suggests that the irrigation-return flows have not yet travelled through the thick unsaturated zone (more than 200 ft thick) in the Mainside subbasin. However, water-level data beneath the MCAGCC golf course in the Mesquite subbasin suggest that irrigation-return flows reached the water table in the early 2000s (fig. 11). For the purposes of this study, irrigation-return flows were not considered a significant source of water and were not simulated.

Groundwater Underflow

Recharge as groundwater underflow across the Emerson, Copper Mountain, and Pinto Mountain Faults was simulated using specified-flux boundary conditions. The fluxes were simulated using the WEL package (McDonald and Harbaugh, 1988), which simulates constant rates of discharge or recharge per stress period at user-selected model cells. Groundwater underflow along the Emerson fault was estimated to be 128 acre-ft/yr according to the model results of Londquist and Martin (1991), and groundwater flow along the Copper Mountain Fault was estimated to be 207 acre-ft/yr according to the model results of Nishikawa and others (2004). Initial rates of groundwater underflow along the Pinto Mountain Fault were estimated from a water budget developed for the Mesquite subbasin that was described in the "Natural Recharge and Discharge" section of this report. The groundwater underflow across Pinto Mountain Fault was separated into two parts and placed in the cells where the Mesquite Lake Wash and the Twentynine Palms Channel cross the southern boundary of the regional model (fig. 17). The distribution and rate of the groundwater underflow across the western and southern boundaries of the regional model were modified during the calibration process. Both estimated and simulated groundwater underflow recharge are listed in table 6.

Infiltration of Runoff

Recharge from the infiltration of runoff in streams within the active model domain was estimated using the INFILv3 watershed model developed for the Joshua Tree area (Nishikawa and others, 2004). The INFILv3 watershed model is a precipitation-runoff model that provides continuous daily simulation of surface and shallow sub-surface water balance (Hevesi and others, 2003). The INFILv3 watershed model covers both the Joshua Tree and the Twentynine Palms areas; however, only the results for the Joshua Tree area were published in a report by Nishikawa and others (2004). The

Table 3. Calibrated hydraulic-conductivity, specific-storage, and specific-yield values of the regional groundwater-flow model of the Twentynine Palms area, California.

[ft/d, foot per day; ft, foot; ft³, foot cubed;

| Subbasin | Zone | Hydraulic conductivity (ft/d) | | | | | | Storage coefficient | | | | | |
| | | Horizontal | | | Vertical | | | Specific storage (1/ft) | | | Specific yield (ft³/ft³) | | |
		Layer 1	Layer 2	Layer 3	Layer 1	Layer 2	Layer 3	Layer 1	Layer 2	Layer 3	Layer 1	Layer 2	Layer 3
Surprise Spring	Outer	48	5	0.20	0.48	0.05	0.0020	3.00E-06	3.00E-06	1.00E-06	0.25	0.16	0.05
	Middle	55	6	.25	.55	.06	.0025	3.00E-06	3.00E-06	1.00E-06	.25	.16	.05
	Inner	[1]73	8	.33	[1]73	.08	.0033	3.00E-06	3.00E-06	1.00E-06	[1].25	.16	.05
		[2]55	8	.33	[2]55	.08	.0033	3.00E-06	3.00E-06	1.00E-06	[2].16	.16	.05
Deadman	West and middle	Dry	20	.20	Dry	.20	.0020	3.00E-06	3.00E-06	1.00E-06	.25	.16	.05
	East	Dry	4	.03	Dry	.04	.0003	3.00E-06	3.00E-06	1.00E-06	.25	.16	.05
Mesquite	West and middle	Dry	24	.20	Dry	.24	.0020	3.00E-06	3.00E-06	1.00E-06	.25	.16	.05
	East	44	24	.20	.44	.24	.0020	3.00E-06	3.00E-06	1.00E-06	.25	.16	.05
Mainside		Dry	4	.03	Dry	.04	.0003	3.00E-06	3.00E-06	1.00E-06	.25	.16	.05

[1] Calibrated hydraulic parameter used in model period a (1952–2000).

[2] Calibrated hydraulic parameter used in model period b (2001–2007) and water-management scenarios (2008–2017).

Table 4. Calibrated transmissivity values of the regional groundwater-flow model of the Twentynine Palms area, California.

[The listed transmissivity value represents the average transmissivity of each hydrogeological zone for the predevelopment condition. ft, foot; ft²/d, foot squared per day; ft/d, foot per day]

Layer	Aquifer	Average highest water level in zone (ft)	Altitude of layer bottom (ft)	Maximum saturated thickness (ft) to model layer bottom	Calibrated transmissivity (ft²/d)	Calibrated hydraulic conductivity (ft/d)
			Surprise Spring Subbasin			
			Inner zone			
1	Upper	2,250	1,880	370	29,352	72.60
2	Middle		1,580	300		8.30
			Middle zone			
1	Upper	2,250	1,980	270	16,992	55.00
2	Middle		1,640	340		6.30
			Outer zone			
1	Upper	2,250	2,200	50	4,510	48.40
2	Middle		1,820	380		5.50
			Mesquite Subbasin			
			West and middle zones			
1	Upper	1,860	1,880	0	7,050	Dry
2	Middle		1,560	300		23.50
			East zone			
1	Upper	1,780	1,740	40	9,750	44.00
2	Middle		1,400	340		23.50
			Deadman Subbasin			
			West and middle zone			
1	Upper	1,830	1,880	0	5,400	Dry
2	Middle		1,560	270		20.00
			East zone			
1	Upper	1,800	1,820	0	1,200	Dry
2	Middle		1,500	300		4.00
			Mainside Subbasin			
1	Upper	1,550	1,680	0	1,400	Dry
2	Middle		1,200	350		4.00

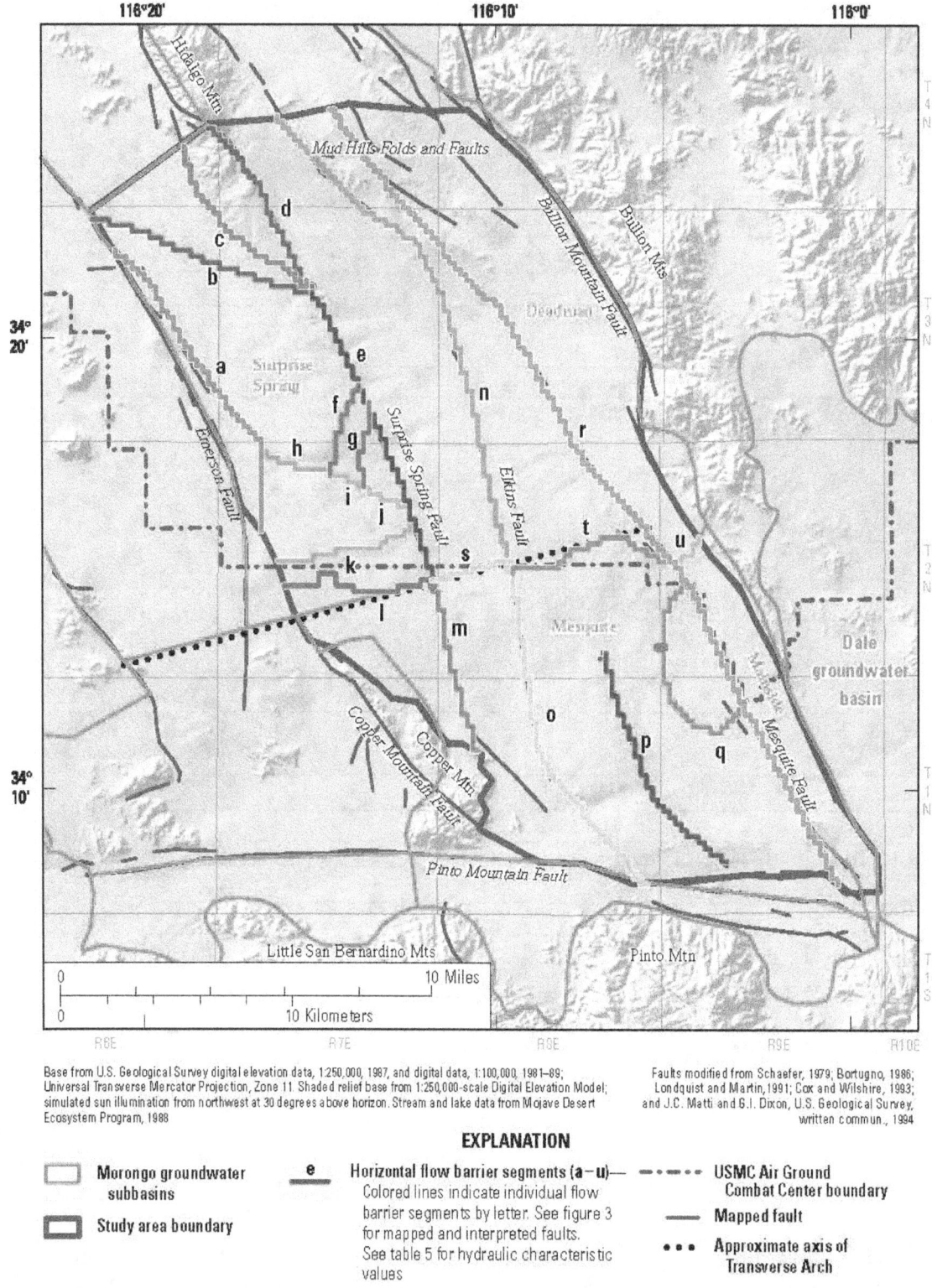

Base from U.S. Geological Survey digital elevation data, 1:250,000, 1997, and digital data, 1:100,000, 1981–89; Universal Transverse Mercator Projection, Zone 11. Shaded relief base from 1:250,000-scale Digital Elevation Model; simulated sun illumination from northwest at 30 degrees above horizon. Stream and lake data from Mojave Desert Ecosystem Program, 1988

Faults modified from Schaefer, 1979; Bortugno, 1986; Londquist and Martin, 1991; Cox and Wilshire, 1993; and J.C. Matti and G.I. Dixon, U.S. Geological Survey, written commun., 1994

EXPLANATION

☐ Morongo groundwater subbasins

▭ Study area boundary

e ▬ Horizontal flow barrier segments (a–u)— Colored lines indicate individual flow barrier segments by letter. See figure 3 for mapped and interpreted faults. See table 5 for hydraulic characteristic values

–·–··– USMC Air Ground Combat Center boundary

—— Mapped fault

••• Approximate axis of Transverse Arch

Figure 19. Location of simulated horizontal-flow barriers for the regional groundwater-flow model of the Twentynine Palms area, California.

Table 5.   Calibrated hydraulic-characteristic, drain-conductance, and general-head boundary conductance values of the regional groundwater-flow model of the Twentynine Palms area, California.

[**Abbreviations:** HFB, horizontal flow barrier; GHB, general-head boundary; NA, not available. Hydraulic characteristic values are in per day. Conductance values of drains and GHBs are in square feet per day. See *figure 19* for HFB locations and *figure 17* for GHB and drain locations]

Hydraulic characteristic				
HFB segment	Description of location	Layer 1	Layer 2	Layer 3
a	Surprise Spring subbasin	1.15E−04	1.15E−04	1.15E−04
b	Surprise Spring subbasin	3.28E−01	3.28E−01	3.28E−01
c	Surprise Spring subbasin	Boundary		
d	Upper Surprise Spring Fault	Boundary		
e	Middle Surprise Spring Fault	1.64E−06	2.13E−06	1.64E−06
f	Surprise Spring subbasin	1.05E−03	1.15E−03	2.30E−04
g	Surprise Spring subbasin	[1] 9.84E−04	3.93E−05	3.93E−05
		[2] 3.28E−05		
h	Surprise Spring subbasin	3.93E−05	3.93E−05	3.93E−05
i	Surprise Spring subbasin	1.05E−03	2.30E−02	2.30E−04
j	Surprise Spring subbasin	9.84E−04	7.87E−04	3.93E−05
k	Surprise Spring subbasin	1.21E−05	1.21E−05	1.21E−05
l	Transverse Arch	3.28E−11	3.28E−11	3.28E−11
m	Lower Surprise Spring Fault	1.64E−06	4.59E−05	6.56E−05
n	Upper Elkin Fault	5.58E−06	1.12E−05	5.58E−06
o	Lower Elkin Fault	6.56E−05	1.90E−04	6.56E−07
p	Mesquite subbasin	3.28E−04	3.28E−04	3.28E−04
q	Mesquite subbasin	2.62E−03	2.62E−03	2.62E−03
r	Mesquite Fault	1.31E−07	1.31E−07	1.31E−07
s	Transverse Arch	3.28E−06	3.28E−06	3.28E−06
t	Transverse Arch	8.20E−06	8.20E−06	8.20E−06
u	Transverse Arch	3.28E−05	3.28E−05	3.28E−05
Drain conductance				
Drain location	Description of location	Layer 1	Layer 2	Layer 3
Surprise Spring	Surprise Spring subbasin	6.46E+01	NA	NA
Deadman Lake	Deadman subbasin	6.67E+02	NA	NA
Mesquite Lake	Deadman and Mesquite subbasins	6.67E+02	NA	NA
General-head boundary conductance				
GHB location	Description of location	Layer 1	Layer 2	Layer 3
Southeast Mainside subbasin	Mainside subbasin and Dale groundwater basin	NA	1.29E+03	NA

[1] Calibrated hydraulic parameter used in model period a (1952–2000).

[2] Calibrated hydraulic parameter used in model period b (2001–2007) and water-management scenarios (2008–2017).

Table 6. Estimated and simulated water-budget components for steady-state conditions of the regional groundwater-flow model of the Twentynine Palms area, California.

[All values are in acre-feet per year]

Recharge	Groundwater underflow across Emerson Fault	Groundwater underflow across Copper Mountain Fault	Groundwater underflow across Pinto Mountain Fault		Stream infiltration	Total
Subbasin	Surprise Spring	Mesquite	Mesquite		Mesquite	
			Mesquite Lake Wash	Twentynine Palms Channel		
Estimated	128	207	255	255	165	1,010
Simulated	110	207	332	180	152	981

Discharge	Spring discharge	Evaporation of moist soils		Evapotranspiration			Groundwater underflow to Dale groundwater basin	Total
Subbasin	Surprise Spring	Deadman	Mesquite	Surprise Spring	Deadman	Mesquite	Mainside	
Estimated	15	0	440	75	30	450	0	1,010
Simulated	7	0	399	59	33	447	36	981

INFILv3 simulated average annual recharge from the infiltration of runoff in the active model domain for 1950 through 1999 was about 165 acre-ft/yr, with essentially all of the recharge simulated in the Mesquite subbasin (Joseph Hevesi, U.S. Geological Survey, written commun., 2008) (table 6). The rate and location of recharge from the infiltration of runoff was specified as a distributed constant flux using the RCH package (McDonald and Harbaugh, 1988) in the groundwater-flow model (fig. 17). The individual model cell recharge rates and distribution were modified slightly during model calibration. The calibrated rate of recharge from the infiltration of runoff was 152 acre-ft/yr, slightly lower than the INFILv3 results (table 6).

Simulated Discharge

Natural groundwater discharge from the model domain consists of flow at the Surprise Spring, evaporation of moisture from wet soil on and surrounding the Mesquite Lake (dry) playa, transpiration of phreatophytes in the Surprise Spring, Deadman, and Mesquite subbasins, and groundwater underflow across the southeastern boundary of the Mainside subbasin. Since 1953, groundwater pumpage, primarily by MCAGCC in the Surprise Spring subbasin, has become the main discharge from the model domain.

Spring Flow

Discharge at the Surprise Spring was estimated to be 15 acre-ft/yr before groundwater development in the Surprise Spring subbasin and ceased shortly after groundwater development began in 1953 (Riley and Worts, 1953 [2001]). Discharge from the spring was simulated by a drain boundary condition (DRN package [McDonald and Harbaugh, 1988])

at the location of the Surprise Spring (yellow cell adjacent to well 35P2 in fig. 17,). The drain boundary is a head-dependent flux boundary, which discharges water from the model domain at a rate proportional to the difference between the simulated head in the drain cell and a specified altitude. The drain cell discharges water only as long as the simulated head is above the specified altitude. The drain altitude was set to the land-surface altitude at the Surprise Spring. The constant of proportionality (drain conductance) was estimated to be 64.6 ft^2/d during the calibration process (table 5). The calibrated steady-state discharge from the spring was 7 acre-ft/yr (table 6).

Evaporation from the Playa Surface and Soil Adjacent to the Playa

Discharge in the form of evaporation from the playa surface of Mesquite Lake (dry) was estimated to be 100 acre-ft/yr before significant groundwater development in the Mesquite subbasin (Riley and Worts, 1953 [2001]). In addition, about 340 acre-ft/yr was estimated to evaporate from moist soil adjacent to the playa surface (Riley and Worts, 1953 [2001]). The combination of these two discharges (440 acre-ft/yr) at and near Mesquite Lake (dry) was simulated using the drain (DRN) package (McDonald and Harbaugh, 1988). Drain cells were specified to approximate the 400-acre barren playa surface and the 340-acre seasonally moist soil adjacent to the playa (fig. 17). All drains were assigned a uniform altitude and drain conductance value. The drain altitude was set equal to the playa surface. The drain conductance value was estimated to be 667 ft^2/d during model calibration (table 5). The calibrated steady-state evaporation simulated by drain cells was about 399 acre-ft/yr (table 6).

Surface discharge by evaporation from the Deadman Lake (dry) playa was assumed to be negligible (Riley and Worts, 1953 [2001]). Drain cells were specified to approximate the Deadman Lake (dry) playa (figs. 2, 17). All drain cells were assigned a uniform altitude and a drain conductance value. The drain altitude was set equal to the altitude of the playa surface, and the drain conductance value was assumed to be equivalent to the drain conductance value calibrated for the playa surface for Mesquite Lake (dry) (667 ft²/d). The simulated discharge at the Deadman Lake (dry) drains was zero for steady-state and transient simulations, which is consistent with Riley and Worts' (1953 [2001]) observation.

Transpiration by Phreatophytes

Transpiration by phreatophytes near the Surprise Spring, Deadman Lake (dry), and Mesquite Lake (dry) was estimated to be 75, 30, and 450 acre-ft/yr, respectively, before groundwater development (Riley and Worts, 1953 [2001]). Transpiration by phreatophytes was simulated in the groundwater-flow model using the Evapotranspiration (EVT) package (McDonald and Harbaugh, 1988). EVT cells were placed in areas where phreatophytes were mapped by Riley and Worts (1953 [2001]) (fig. 17). The evapotranspiration rate was assumed to be at a maximum when the water table was at land surface and to decrease linearly to zero when the water table was 50 ft below land surface. The extinction depth of 50 ft represents an average depth for deep-rooted mesquite. The maximum evapotranspiration rate was set to be 0.7 ft/yr in Surprise Spring and Deadman subbasins and 11.25 ft/yr in the Mesquite subbasin on the basis of phreatophyte type and density in each subbasin. The simulated evapotranspiration for steady-state conditions was 59, 33, and 447acre-ft/yr for the Surprise Spring, Deadman, and Mesquite subbasins, respectively (table 6).

Groundwater Underflow

Discharge as groundwater underflow across the southeast boundary of Mainside subbasin was simulated using GHB cells in model-layer 2, where the Twentynine Palms Channel crosses the eastern boundary of the Mainside subbasin (fig. 17). The GHB was placed in model layer 2 because model-layer 1 is dry in that area of the model domain. The hydraulic head for the GHB was set to 1,535 ft, about 15 ft lower than water levels in the Mainside subbasin, and was constant during the simulation. GHB conductance values were initially set to large values, allowing groundwater to flow freely across the modeled boundary. During the calibration process, the conductance values were lowered such that the simulated hydraulic heads in the Mainside subbasin closely matched measured water levels. The simulated GHB conductance was 1290 ft²/d and groundwater outflow was 36 acre-ft/yr for the steady-state simulation (table 5, 6).

Pumpage

Pumpage was simulated using the MNW (Multi-Node Well) package (Halford and Hanson, 2002). The MNW package simulates wells that are completed in multiple aquifers and allows vertical groundwater movement through the well bores. The pumpage is distributed dynamically into model layers (multi-well nodes) on the basis of the transmissivity and simulated hydraulic head associated with each layer during each stress period.

Pumpage from 11 potable water-supply wells in Surprise Spring subbasin and one nonpotable well (3N/8E–29L1) in the Deadman subbasin were metered and reported by the MCAGCC (figs.10, 17). For 1953 through 1983, only annual pumpage data were available. Pumpage data reported by the MCAGCC were used directly for this period. Monthly pumpage data were available for 1985 through September 2007 for the wells in the Surprise Spring subbasin. Pumpage for October, November, and December of 2007 was assumed to be the same as that recorded for those months in 2006. Missing monthly data from September 1997 and March 2001were estimated by averaging the pumpage from the same well during the same months during the previous and following years.

Pumpage from the Mesquite subbasin included pumpage from a supplemental irrigation-supply well at the MCAGCC (referred to as the MCAGCC golf course well) and from a City of Twentynine Palms supply well (1N/9E–21H1) (fig. 17). Pumpage from the MCAGCC golf course well was not simulated because records are incomplete and extraction was small. Approximately 850 acre-ft/yr of pumpage was reported and simulated for the City of Twentynine Palms production well for the period of operation (March 2003 through September 2007).

Model Calibration

The regional groundwater-flow model was calibrated using a trial-and-error approach; the estimates of the aquifer properties and the groundwater underflows from adjacent subbasins were iteratively adjusted to improve the match between simulated hydraulic heads and measured groundwater levels. Measured groundwater levels for the steady state (before 1953) and transient conditions (1953 through 2007) were used to calibrate the model. The model was calibrated in an iterative manner between the steady-state and transient simulations. Starting with the steady-state model, an initial hydraulic-head distribution for pre-development conditions was simulated by adjusting groundwater underflow, hydraulic conductivity, hydraulic-characteristic values of the horizontal flow barriers (faults), and conductances of head-dependent boundaries. The simulated head distribution from the steady-state model was then used as the initial-head distribution for the transient model. Initial estimates of aquifer storage properties were refined during the transient simulation. If a satisfactory match between measured and simulated results was not obtained, the process was repeated. Initial estimates of model parameters were adjusted during calibration within limits that were set based on the geologic and hydrologic characteristics of the basin and the degree of confidence placed on the original data estimates.

Calibration of Steady-State Conditions

A steady-state condition reflects a system in equilibrium: the recharge and discharge are equal, and the hydraulic heads and the volume of water stored within the system do not change. Steady-state hydraulic heads depend on the quantities and the distribution of recharge to and discharge from the groundwater system, and on the aquifer properties that control the groundwater flow as it moves through the aquifer system. Model parameters representing these properties include the horizontal and vertical conductivity, hydraulic-characteristic values of the horizontal-flow barriers (faults), and the conductance of drain and general-head boundaries. These parameters were adjusted along with groundwater underflow and maximum rate of evapotranspiration to match simulated steady-state hydraulic heads with measured water levels at selected observation points (or calibration wells) across the three model layers and four subbasins.

The highest water level measured at each monitoring well in the study area was assumed to represent predevelopment or steady-state conditions. This assumption was necessary because of the absence of water-level data for most of the model domain before 1953. The assumption probably is reasonable for this study area because data from most areas in the model domain show no significant water-level change during the period of record. Water-level data for monitoring wells in the Surprise Spring subbasin were carefully screened on the basis of the measurement date and the possibility of being influenced by pumping in the subbasin. A total of 20 wells with water-level measurements were chosen as observation points for the model: 6 in layer 1, 12 in layer 2, and 2 in layer 3 (fig. 20).

During the steady-state calibration, the horizontal hydraulic conductivity values were modified from initial estimates until simulated hydraulic heads were close to measured steady-state or predevelopment water levels (fig. 20). Initial estimates of hydraulic conductivity values were derived from specific-capacity tests (table 1) and results from previous models: a two-layer model of the Surprise Spring subbasin (table 2; Londquist and Martin, 1991), and a three-layer model of the Joshua Tree area (table 2; Nishikawa and others, 2004). The final calibrated hydraulic conductivity values of all layers are similar to those of the Joshua Tree model except for layer 2 (middle aquifer) in the Mesquite subbasin and the middle and west hydrogeologic zones of the Deadman subbasin (tables 2, 3; fig. 18). The calibrated hydraulic conductivity values for layer 2 in these hydrogeologic zones are about five times higher than those calibrated by Nishikawa and others (2004) for the middle aquifer. The calibrated hydraulic conductivity values for layer 3 are slightly lower than those estimated

for the lower aquifer by Londquist and Martin (1991). Values for the upper two aquifers are not directly comparable with those estimated by Londquist and Martin (1991) because the Surprise Spring model is a two-layer model that simulates the upper and middle aquifers as one model layer.

The calibrated transmissivity values for the combined upper and middle aquifers of the inner and outer zones in the Surprise Spring subbasin (table 4) are close to transmissivity values estimated from specific-capacity tests for wells perforated in the upper and middle aquifers (table 1). The calibrated transmissivity values for the combined upper and middle aquifers of the middle hydrogeologic zone of the Surprise Spring subbasin are more than twice that estimated from specific-capacity values. Specific-capacity tests can underestimate aquifer transmissivity because most wells do not penetrate the entire thickness of the aquifer. For hydrogeological zones in other subbasins, the calibrated transmissivity values are within the range of transmissivity values estimated from specific-capacity tests (tables 1, 4).

The ratio of the vertical hydraulic conductivity for each model layer to the horizontal conductivity of the model layer was set at 1 to 100. The small ratio between vertical and horizontal hydraulic conductivity was used to represent interfingering of coarse and fine-grained sediments in the area. The ratio was not adjusted during the steady-state calibration process.

The simulated steady-state recharge from groundwater underflow and the infiltration of runoff were decreased slightly from estimated values during calibration so that the simulated hydraulic heads would more closely match the measured water levels (fig. 21, table 6). Simulated groundwater underflow across the Emerson Fault was decreased from 128 to 110 acre-ft/yr . The estimated quantity of groundwater underflow across the Pinto Mountain Fault (510 acre-ft/yr) was divided equally between the Mesquite Lake Wash and the Twentynine Palms Channel (fig. 21). The final calibrated groundwater underflow was increased beneath the Mesquite Lake Wash from 255 to 332 acre-ft/yr and decreased beneath the Twentynine Palms Channel from 255 to 180 acre-ft/yr (table 6).

The simulated steady-state discharge (evaporation, evapotranspiration, spring discharge, and groundwater underflow) values were close to the values estimated for Surprise Spring and Deadman, and Mesquite subbasins (fig. 21, table 6). The total simulated groundwater recharge or discharge was 981 acre-ft/yr compared with the estimated value of 1,010 acre-ft/yr. The simulated steady-state hydraulic-head distribution was sensitive to the conductances of the drains and the general-head boundary, and to the hydraulic characteristics of simulated faults. Typically, measured or estimated values were not available for these parameters; therefore, the initial values of these parameters were set large enough to allow unrestricted flow through the aquifer system. The steady-state calibration primarily involved adjusting these parameters (table 5) until simulated hydraulic heads were close to measured water levels (fig. 20) and simulated water-budget terms were similar to those estimated (fig. 21).

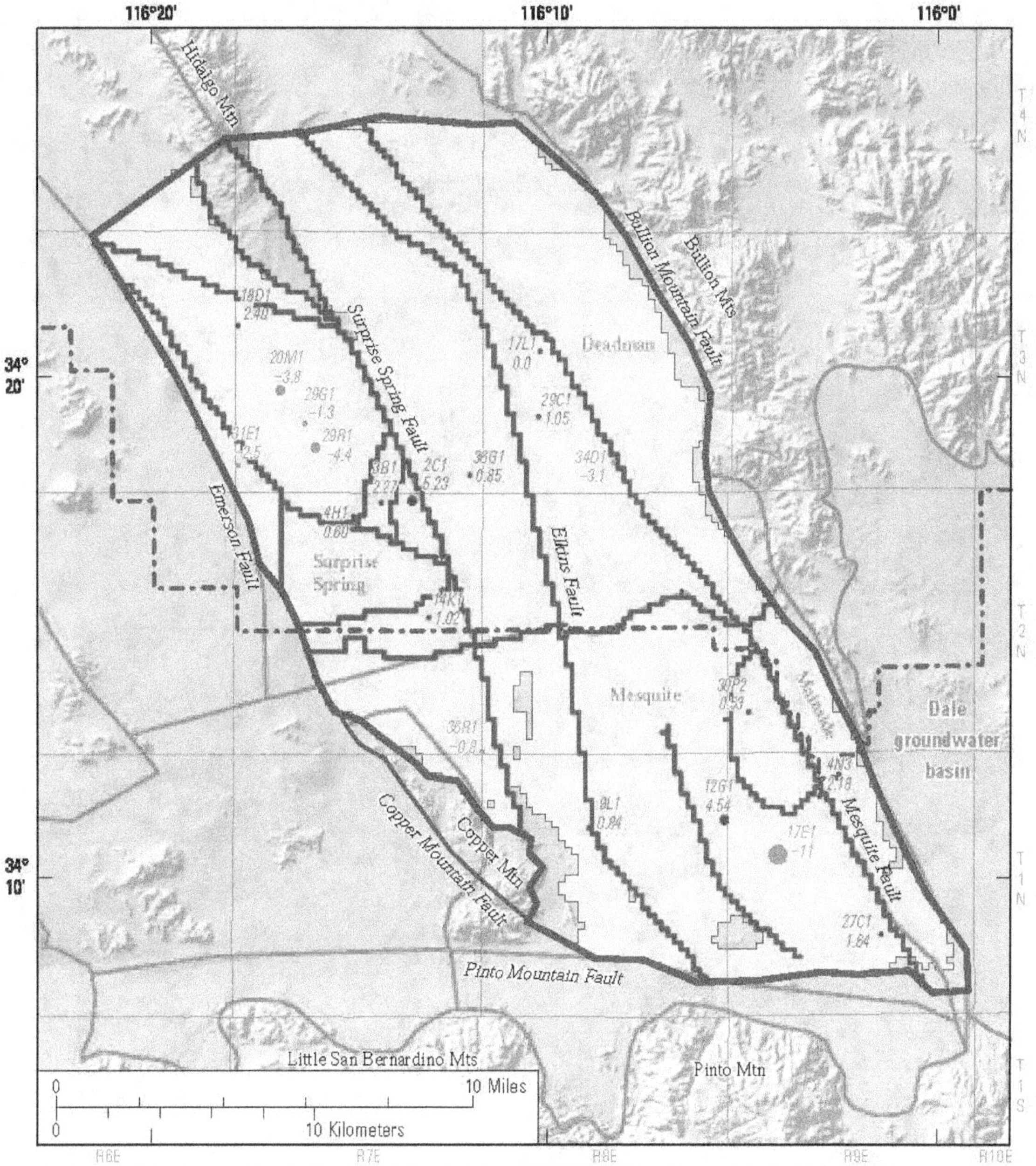

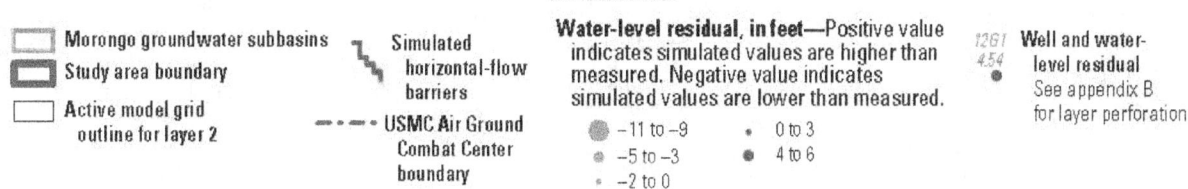

Figure 20. Observation points and water-level residuals for steady-state calibration of the regional groundwater-flow model of the Twentynine Palms area, California.

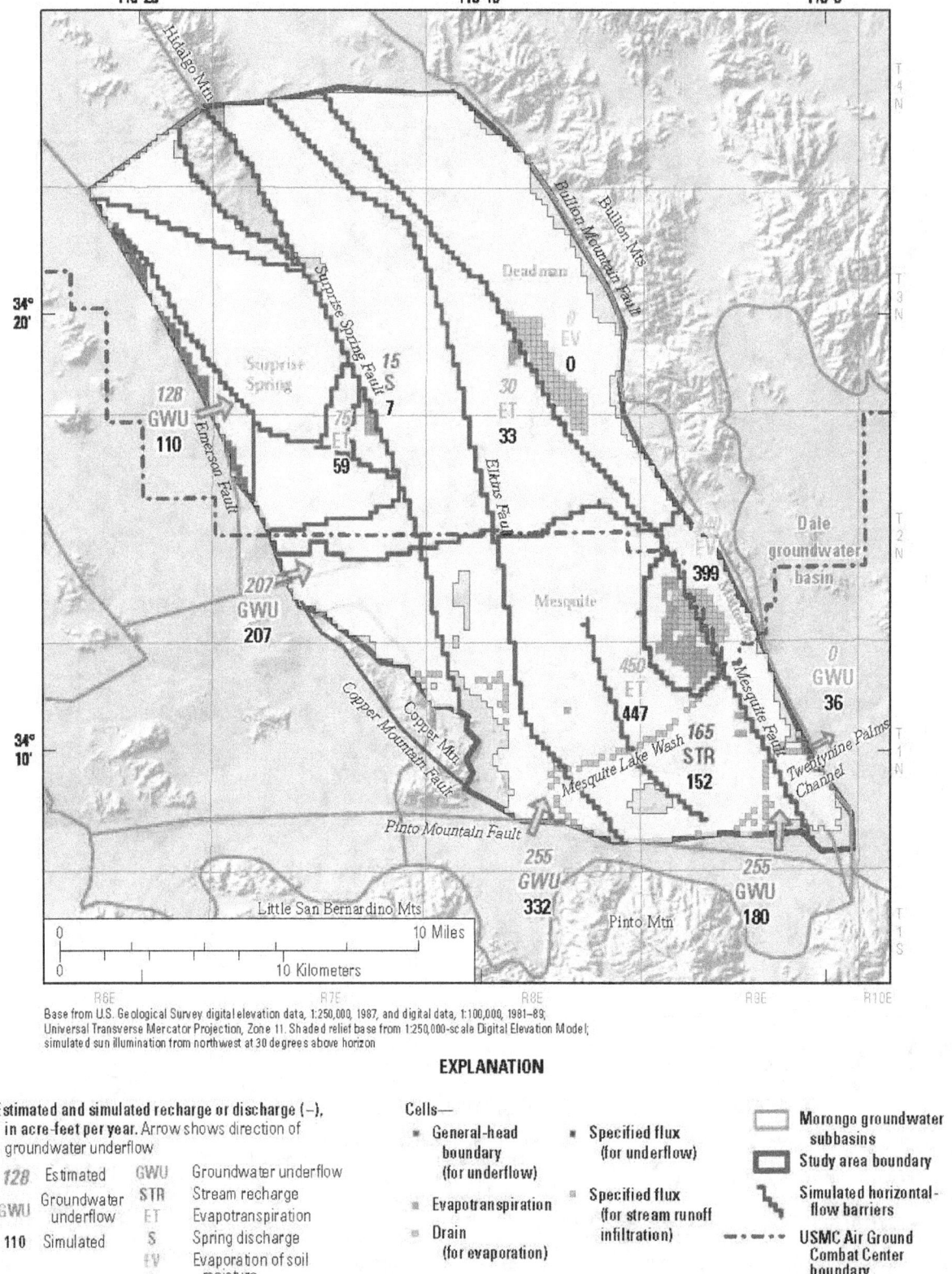

Base from U.S. Geological Survey digital elevation data, 1:250,000, 1987, and digital data, 1:100,000, 1981–89;
Universal Transverse Mercator Projection, Zone 11. Shaded relief base from 1:250,000-scale Digital Elevation Model;
simulated sun illumination from northwest at 30 degrees above horizon

EXPLANATION

Estimated and simulated recharge or discharge (–),
in acre-feet per year. Arrow shows direction of
groundwater underflow

128	Estimated	GWU	Groundwater underflow
GWU	Groundwater underflow	STR	Stream recharge
110	Simulated	ET	Evapotranspiration
		S	Spring discharge
		EV	Evaporation of soil moisture

Cells—
- General-head boundary (for underflow)
- Evapotranspiration
- Drain (for evaporation)
- Specified flux (for underflow)
- Specified flux (for stream runoff infiltration)

- Morongo groundwater subbasins
- Study area boundary
- Simulated horizontal-flow barriers
- USMC Air Ground Combat Center boundary

Figure 21. Estimated and simulated steady-state water-budget components for the Twentynine Palms area, California.

The steady state simulated hydraulic heads were similar to the measured water levels at the observation points. A map showing the difference (residual) between the simulated hydraulic heads and measured water levels at the same observation indicates that only two observation points had an absolute residual greater than 5 feet; the average absolute residual was 2.5 ft (fig. 20). The largest residuals were at well 2N/7E– 2C1 in the Surprise Spring subbasin (5.2 ft) and well 1N/9E– 17E1 in the Mesquite subbasin (-11 ft). The overall root mean square error (RMSE) of the 20 observation points was 3.5 feet and the average error was about 0.25 feet, indicating very little bias. The measured water levels and simulated steady-state hydraulic heads closely follow a 1:1 correlation line (fig. 22). The simulated hydraulic-head contours for the calibrated steady-state groundwater-flow model for model layers 1–3 are shown on figure 23.

Calibration of Transient Conditions

The transient calibration process primarily involved modifying storage properties of the regional model. Simulated hydraulic heads were compared to measured water levels for year 2000 at 23 observation points (fig. 24A) and long-term water-level hydrographs at calibration wells for 1952-2007 (fig. 24B; appendix A). If adjusting storage properties within a reasonable range did not produce a reasonable match between the simulated hydraulic heads and measured water levels, the steady-state model parameters (horizontal and vertical conductivity, hydraulic-characteristic values of the horizontal flow barriers (faults), and the conductance of drain and general-head boundaries) were recalibrated and the transient calibration was reinitiated. This iterative calibration process was completed numerous times until there was a reasonable match between simulated and measured values.

Initial estimates of storage properties (specific storage and specific yield) were based on values calibrated by Londquist and Martin (1991) and Nishikawa and others (2004) for models of the Surprise Spring and Joshua Tree areas, respectively (table 2). The initial estimates of specific storage (1.0×10^{-6} ft^{-1}) (table 2) were increased to 3.0×10^{-6} ft^{-1} in layers 1 and 2 and maintained at 1.0×10^{-6} ft^{-1} in layer 3 (table 3). The transient model was sensitive to variations in specific yield, especially in the parts of the Surprise Spring and Mesquite subbasins where groundwater was pumped. The calibrated specific yield values are similar to those estimated by Londquist and Martin (1991) and Nishikawa and others (2004) (tables 2, 3).

The barrier effect of some faults on groundwater flow was not evident until the aquifer system was stressed by groundwater pumping. For example, most of the faults in the Surprise Spring subbasin were not identified until pumping was initiated in the subbasin (Londquist and Martin, 1991). Consequently, the transient calibration process involved modifying the hydraulic-characteristic values of HFB segments near pumping wells (figs. 17, 19) in order to approximate hydraulic-head declines measured in long-term monitoring

wells (for example, hydrographs for wells 2N/7E–2C1 and 2N/7E–3B1, appendix A, fig. A4). A change in a HFB hydraulic-characteristic value during the transient simulation required an additional iteration of the calibration process, starting with a steady-state simulation using the modified hydraulic-characteristic value, to produce initial conditions for a new transient simulation.

Simulated hydraulic heads for year 2000 generally agreed well with the measured water levels at observation points (figs. 22, 24A). Only two observation points had an absolute residual of more than 5 feet (well 2N/7E–2C1 in the Surprise Spring subbasin and well 1N/9E–17E1 in the Mesquite subbasin) (fig. 24A). These observation points also had a large discrepancy in the steady-state simulation (fig. 20). The average absolute residue of observation points was 2.76 feet. The RMSE was about 4.36 feet and the average error was about -0.24 ft. The simulated hydraulic heads and measured water levels for year 2000 closely follow a 1:1 correlation line (fig. 22).

Simulated water-level hydrographs (or time series of hydraulic heads) for the period of 1953 to 2000 matched closely to the measured hydrographs at model calibration wells (fig. 24 B, appendix A). However, simulated hydraulic heads were higher after 2000 than the measured in values the inner zone of the Surprise Spring subbasin, as shown by the hydrograph of well 2N/7E-2C1 (fig. 24B). Because pumping was metered, the water-level underestimation after year 2000 may be caused by using constant hydraulic properties to represent the heterogeneous upper aquifer of the inner zone. From 1953 to 2000, water levels declined about 120 ft in the inner zone, as indicated by the hydrograph of well 2N/7E-2C1 (fig. 25). Interpretation of the geologic and geophysical logs in the inner zone suggests that the water table in the upper aquifer had dropped into aquifer materials similar to those in the upper aquifer of the adjacent middle zone. Aquifer materials from land surface to a depth of about 120 ft in the upper aquifer of the inner zone are coarser grained and more permeable than those in the lower part of the upper aquifer. Simulating hydraulic heads that approximated measured water levels in the inner zone after 2000 required changing the hydraulic conductivity and specific yield of layer 1 to values calibrated for layer 1 in the middle zone for the simulation period 2000–2007 (table 3, fig. 25). In addition, the hydraulic-characteristic value for HFB segment **g** was reduced to a value similar to the hydraulic characteristic for HFB segment *g* calibrated for layer 2 (table 5).

The modified parameters reasonably match the measured water-level decline in the inner zone of the Surprise Spring subbasin for 2000-2007 (fig. 25). In addition, the simulated 2007 hydraulic-head contours (fig. 26) reasonably match measured 2006 water-table contours presented in Stamos and others (2007) (fig. 12). These results indicate that the model adequately represents historical groundwater conditions in the Surprise Spring, Deadman, Mesquite, and Mainside subbasins.

About 145,450 acre-ft of groundwater pumpage was simulated during the transient simulation period (1953–2007):

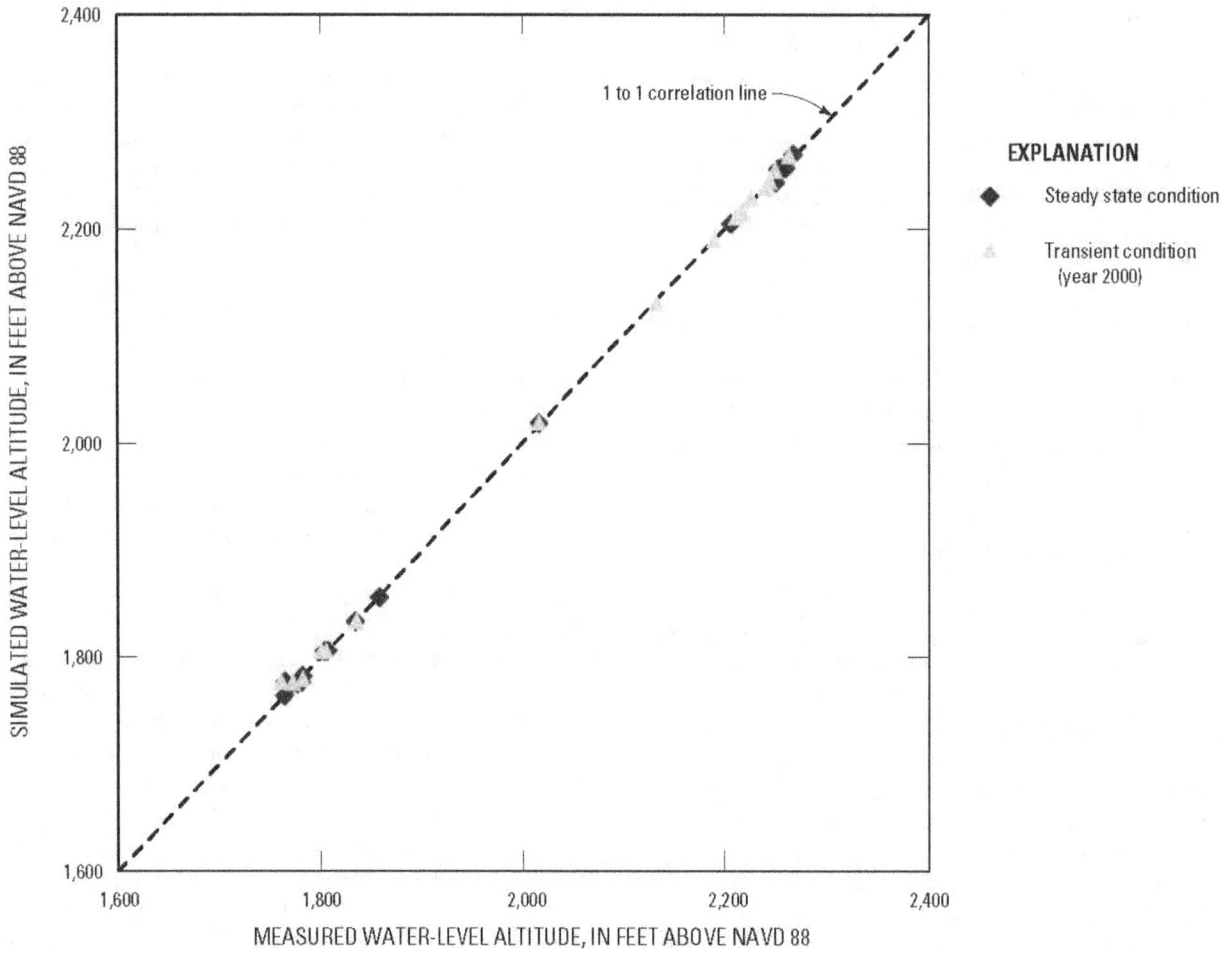

Figure 22.   Relation between simulated hydraulic heads and measured water levels, with 1:1 correlation line for steady-state and transient conditions (year 2000), for the regional groundwater-flow model of the Twentynine Palms area, California.

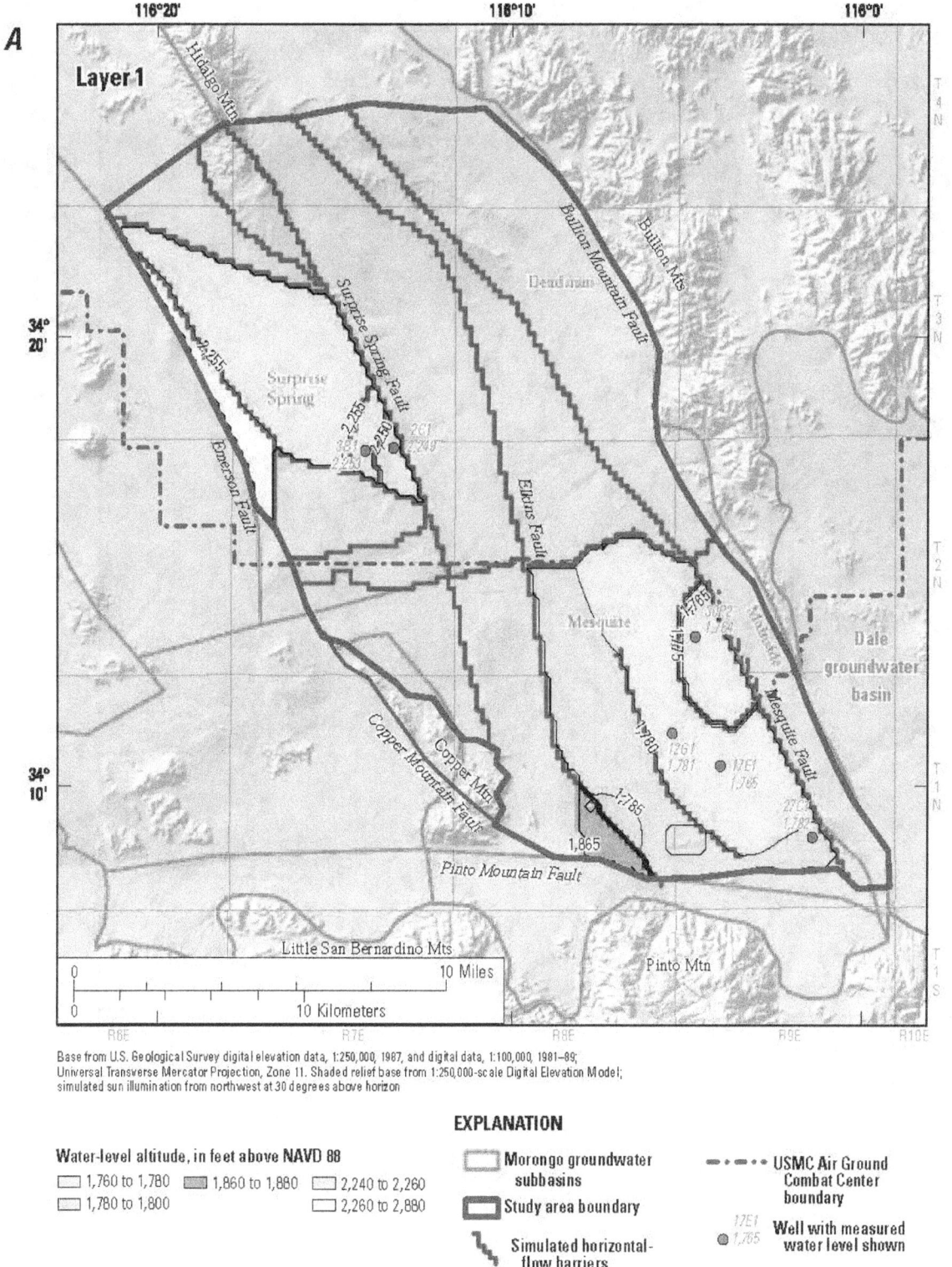

Base from U.S. Geological Survey digital elevation data, 1:250,000, 1987, and digital data, 1:100,000, 1981–89;
Universal Transverse Mercator Projection, Zone 11. Shaded relief base from 1:250,000-scale Digital Elevation Model;
simulated sun illumination from northwest at 30 degrees above horizon

EXPLANATION

Water-level altitude, in feet above NAVD 88

- 1,760 to 1,780
- 1,780 to 1,800
- 1,860 to 1,880
- 2,240 to 2,260
- 2,260 to 2,880

- Morongo groundwater subbasins
- Study area boundary
- Simulated horizontal-flow barriers

—·—·— USMC Air Ground Combat Center boundary

17E1
1,765 Well with measured water level shown

Figure 23. Simulated hydraulic-head contours and measured water levels for steady-state conditions of (A) model layer 1, (B) model layer 2, and (C) model layer 3 of the regional groundwater-flow model of the Twentynine Palms area, California.

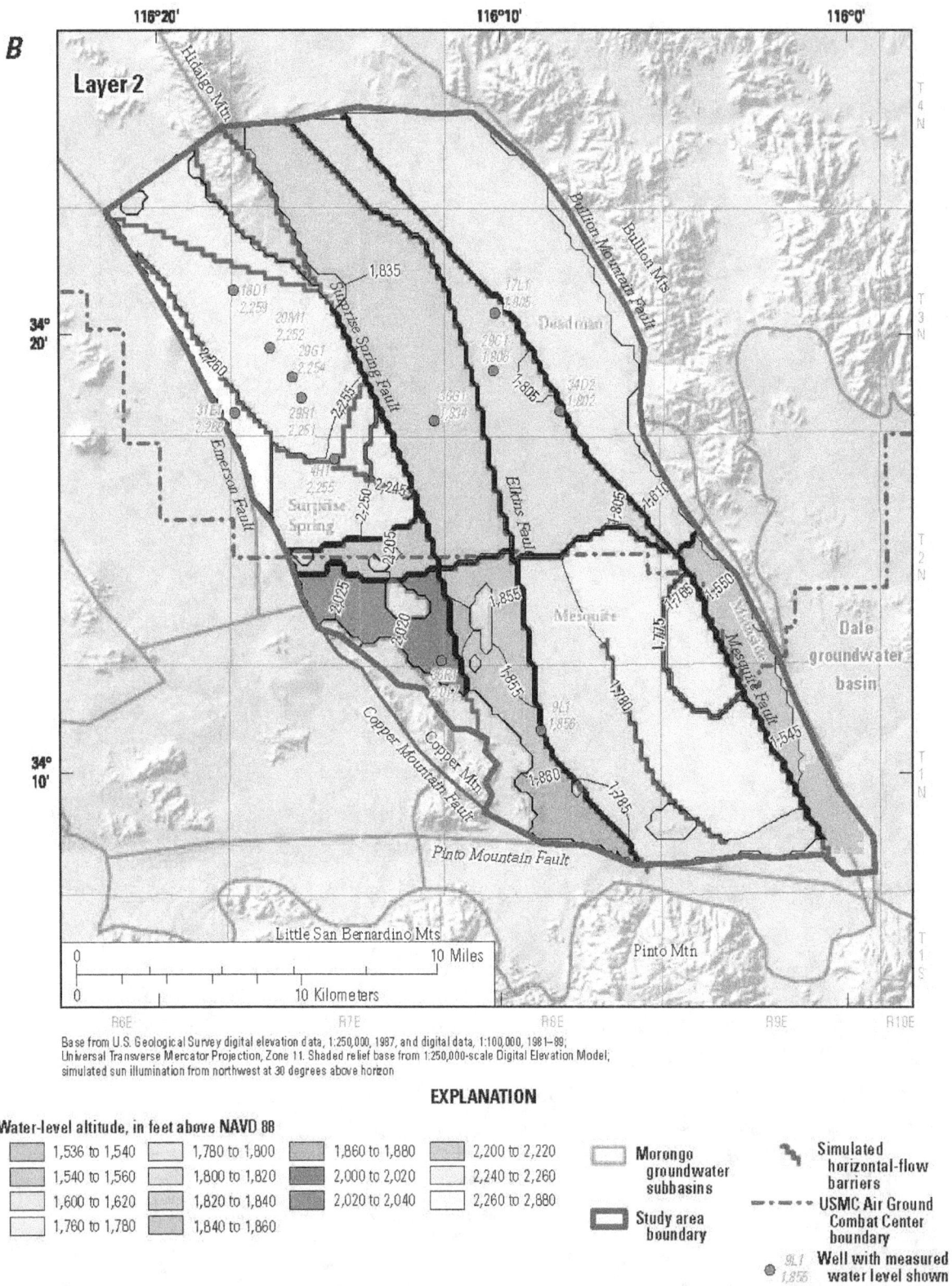

Base from U.S. Geological Survey digital elevation data, 1:250,000, 1987, and digital data, 1:100,000, 1981–89;
Universal Transverse Mercator Projection, Zone 11. Shaded relief base from 1:250,000-scale Digital Elevation Model;
simulated sun illumination from northwest at 30 degrees above horizon

EXPLANATION

Water-level altitude, in feet above NAVD 88

1,536 to 1,540	1,780 to 1,800	1,860 to 1,880	2,200 to 2,220
1,540 to 1,560	1,800 to 1,820	2,000 to 2,020	2,240 to 2,260
1,600 to 1,620	1,820 to 1,840	2,020 to 2,040	2,260 to 2,880
1,760 to 1,780	1,840 to 1,860		

☐ Morongo groundwater subbasins

▭ Study area boundary

⤵ Simulated horizontal-flow barriers

–·–·– USMC Air Ground Combat Center boundary

● 9L1 1,858 Well with measured water level shown

Figure 23.   Continued.

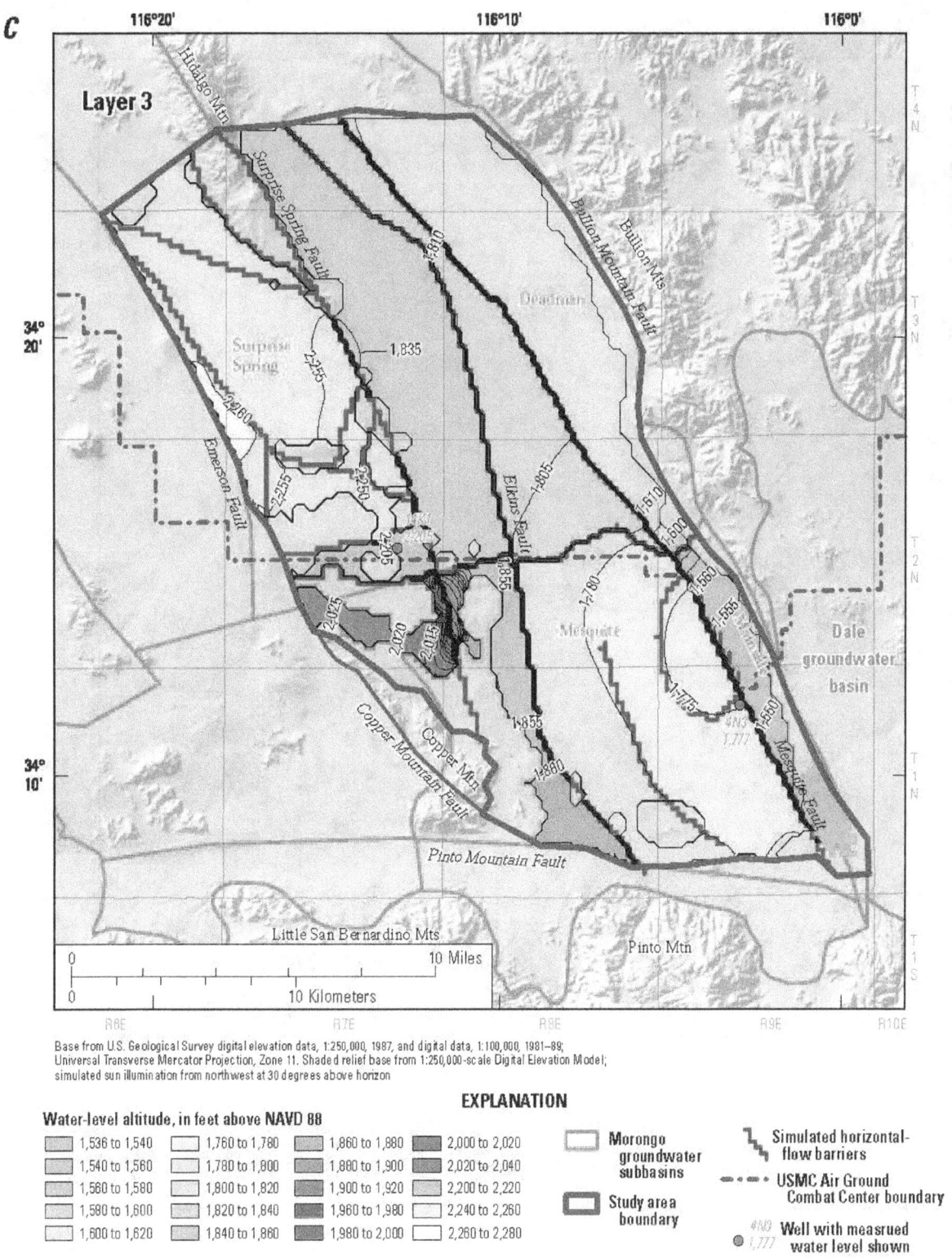

Base from U.S. Geological Survey digital elevation data, 1:250,000, 1987, and digital data, 1:100,000, 1981–89;
Universal Transverse Mercator Projection, Zone 11. Shaded relief base from 1:250,000-scale Digital Elevation Model;
simulated sun illumination from northwest at 30 degrees above horizon

EXPLANATION

Water-level altitude, in feet above NAVD 88

1,536 to 1,540	1,760 to 1,780	1,860 to 1,880	2,000 to 2,020
1,540 to 1,560	1,780 to 1,800	1,880 to 1,900	2,020 to 2,040
1,560 to 1,580	1,800 to 1,820	1,900 to 1,920	2,200 to 2,220
1,580 to 1,600	1,820 to 1,840	1,960 to 1,980	2,240 to 2,260
1,600 to 1,620	1,840 to 1,860	1,980 to 2,000	2,260 to 2,280

Morongo
groundwater
subbasins

Study area
boundary

Simulated horizontal-
flow barriers

USMC Air Ground
Combat Center boundary

4N3 Well with measrued
1,777 water level shown

Figure 23. Continued.

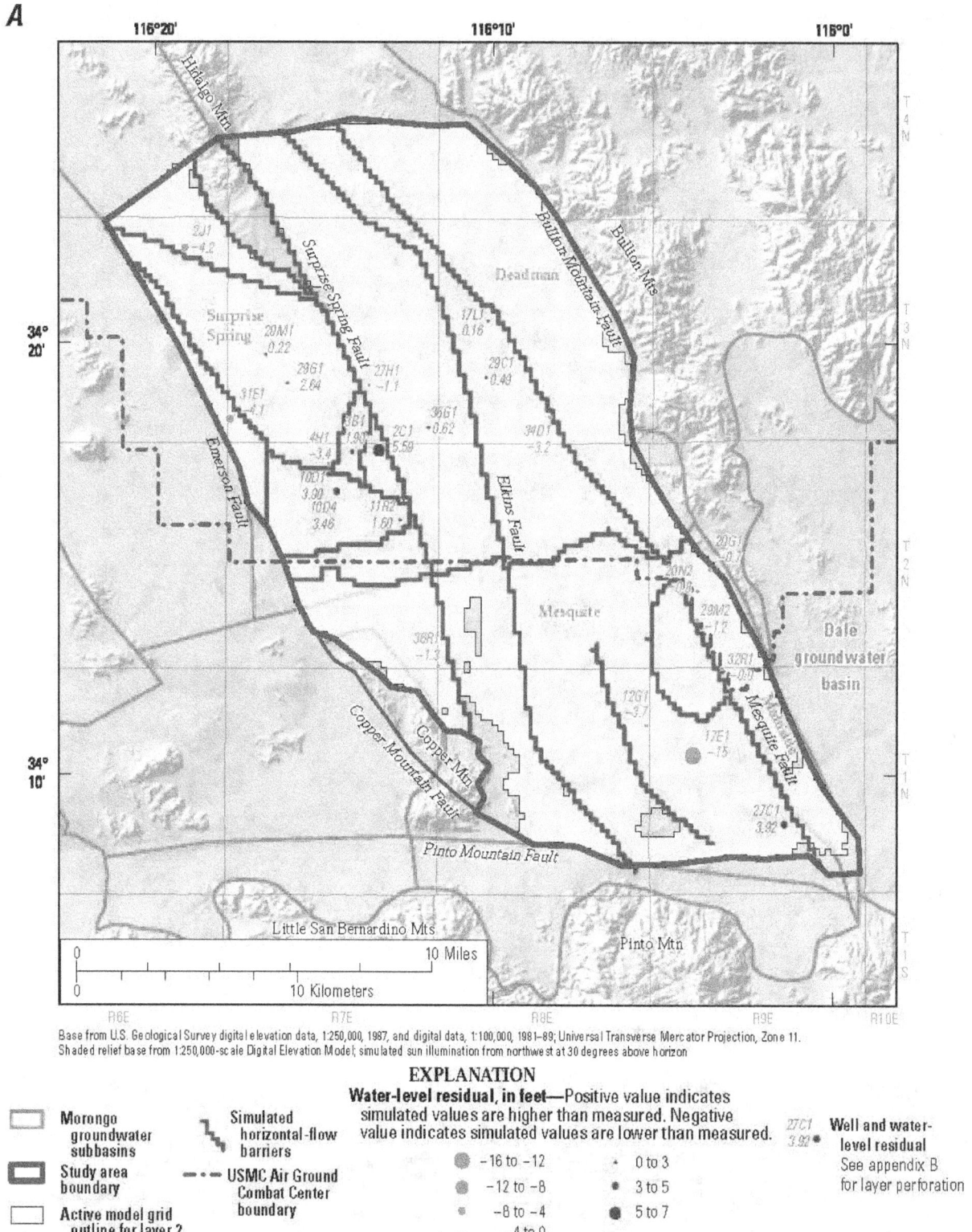

Base from U.S. Geological Survey digital elevation data, 1:250,000, 1987, and digital data, 1:100,000, 1981–89; Universal Transverse Mercator Projection, Zone 11.
Shaded relief base from 1:250,000-scale Digital Elevation Model; simulated sun illumination from northwest at 30 degrees above horizon

EXPLANATION
Water-level residual, in feet—Positive value indicates simulated values are higher than measured. Negative value indicates simulated values are lower than measured.

Morongo groundwater subbasins

Study area boundary

Active model grid outline for layer 2

Simulated horizontal-flow barriers

USMC Air Ground Combat Center boundary

● −16 to −12
● −12 to −8
● −8 to −4
· −4 to 0

· 0 to 3
● 3 to 5
● 5 to 7

27C1
3.92 ● Well and water-level residual
See appendix B for layer perforation

Figure 24.   *(A)* Observation points and residuals between simulated hydraulic heads and measured water levels for the transient simulation (year 2000) of the regional groundwater-flow model and *(B)* representative hydrographs of simulated hydraulic heads and measured water levels (1952–2007) for the Twentynine Palms area, California.

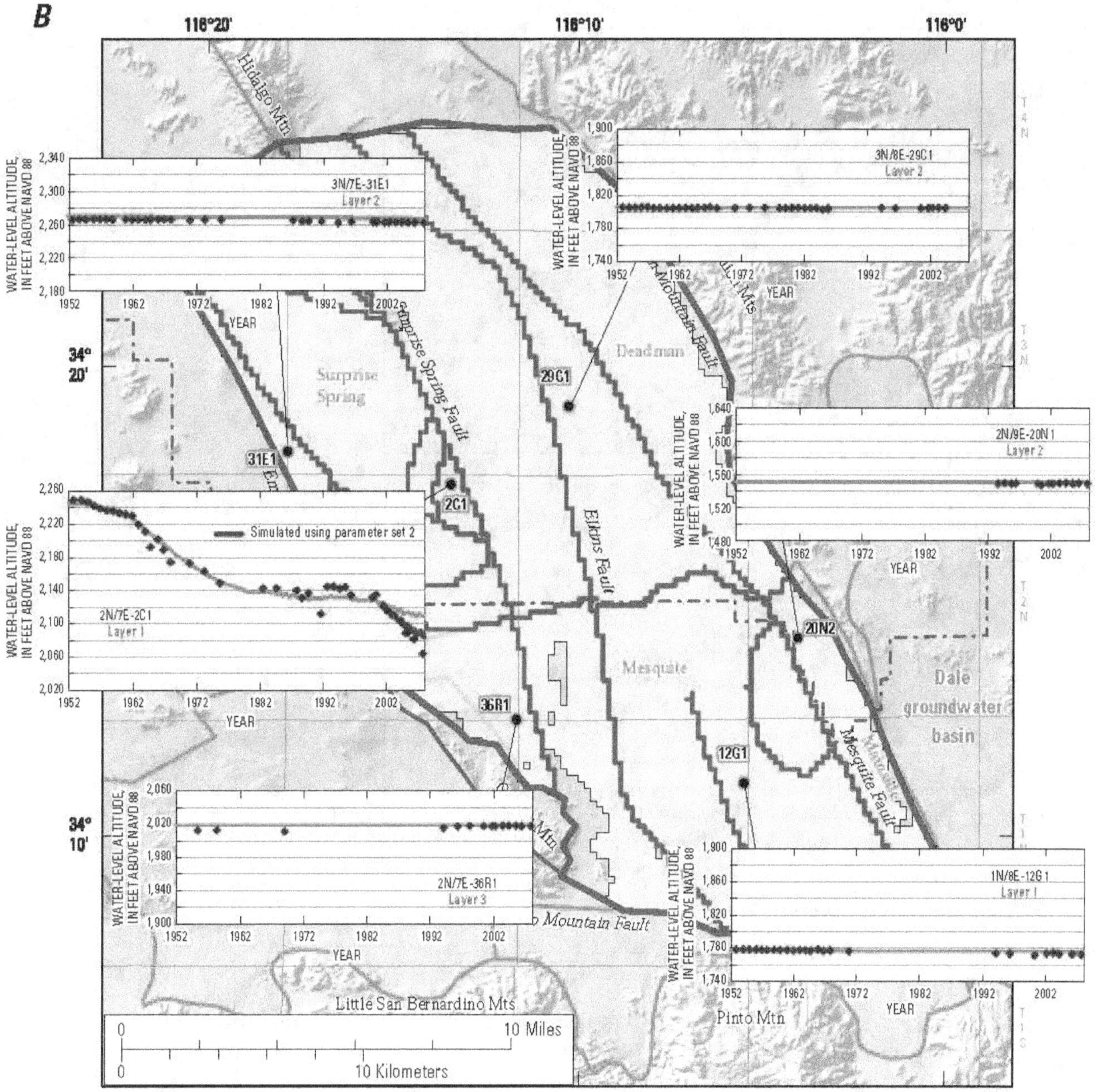

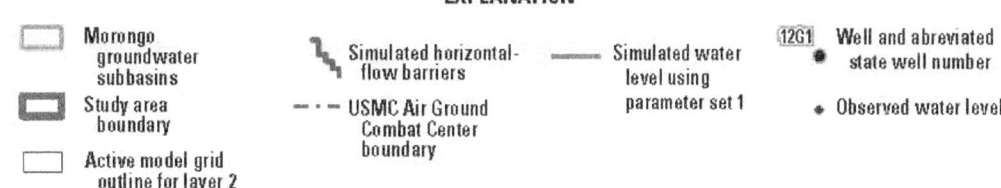

B

Base from U.S. Geological Survey digital elevation data, 1:250,000, 1987, and digital data, 1:100,000, 1981–89; Universal Transverse Mercator Projection, Zone 11.
Shaded relief base from 1:250,000-scale Digital Elevation Model; simulated sun illumination from northwest at 30 degrees above horizon

EXPLANATION

▭	Morongo groundwater subbasins
▭	Study area boundary
▭	Active model grid outline for layer 2

Simulated horizontal-flow barriers

–·–·– USMC Air Ground Combat Center boundary

Simulated water level using parameter set 1

12G1 Well and abreviated state well number

♦ Observed water level

Figure 24. Continued.

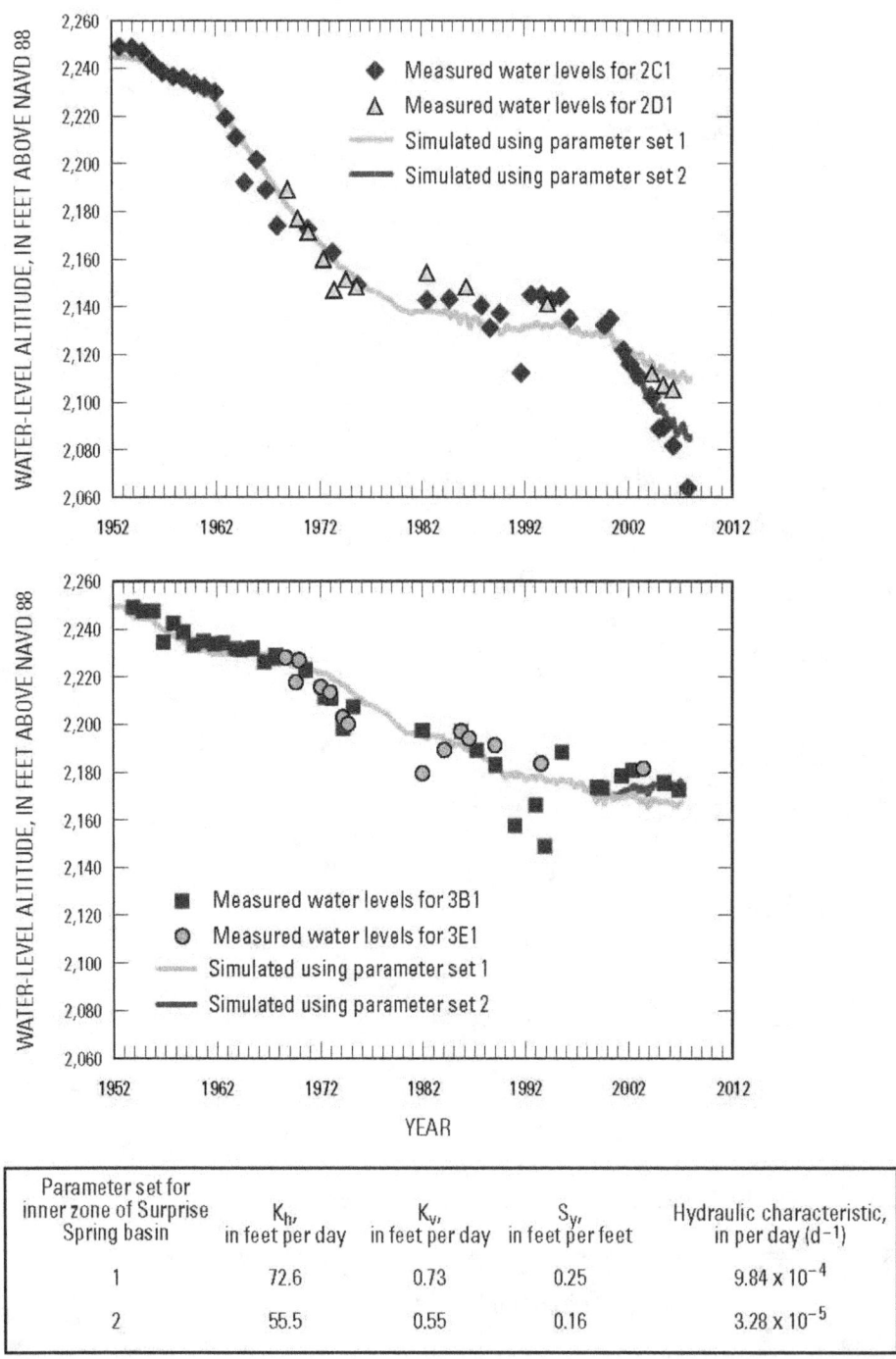

Parameter set for inner zone of Surprise Spring basin	K_h, in feet per day	K_v, in feet per day	S_y, in feet per feet	Hydraulic characteristic, in per day (d^{-1})
1	72.6	0.73	0.25	9.84×10^{-4}
2	55.5	0.55	0.16	3.28×10^{-5}

Figure 25.   Measured water levels and simulated hydraulic heads in wells 2N/7E–2C1, 2D1, 3B1, and 3E1 resulting from the use of different parameter sets for the inner zone of Surprise Spring subbasin of the regional groundwater-flow model of the Twentynine Palms area, California.

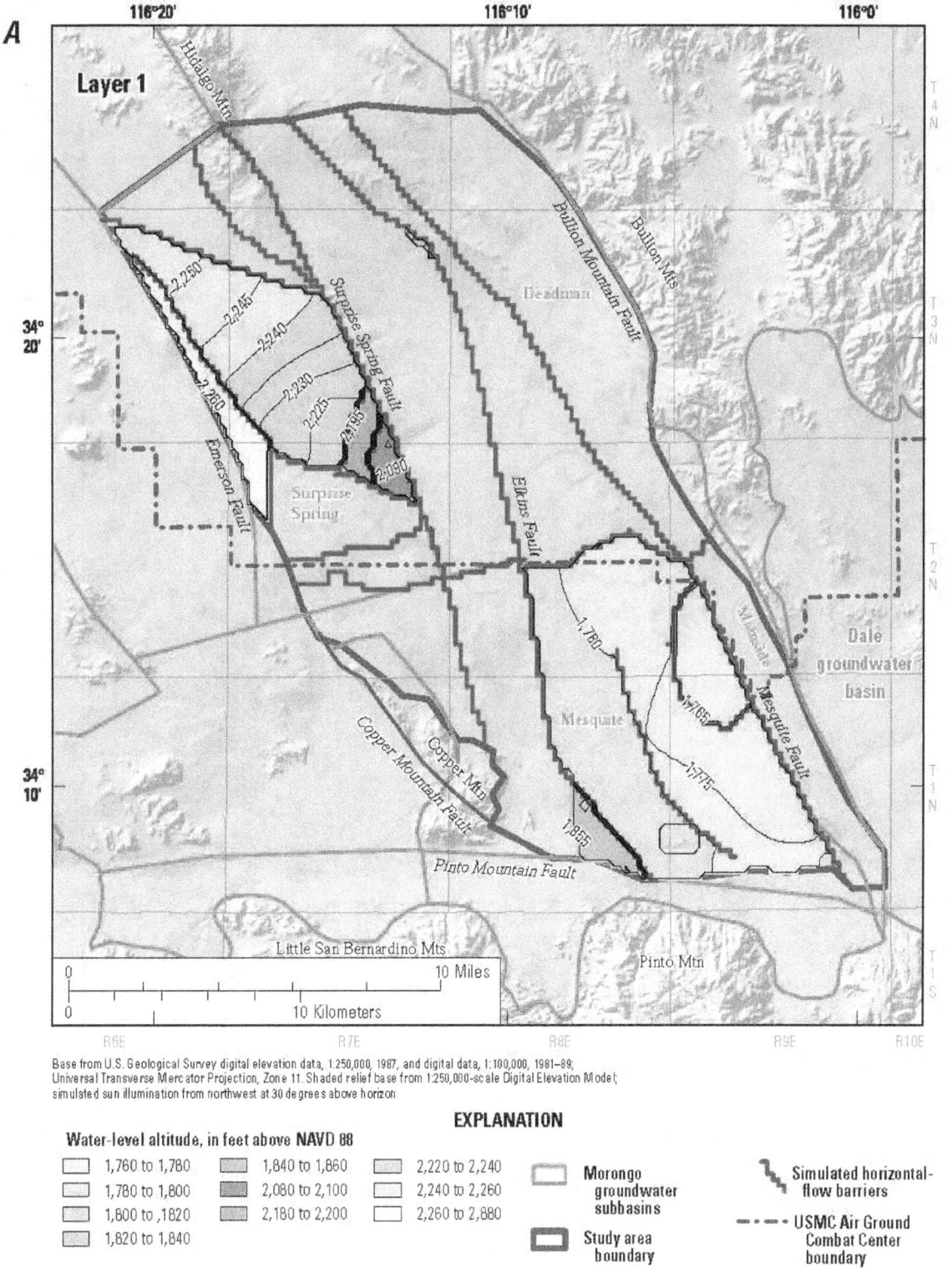

Base from U.S. Geological Survey digital elevation data, 1:250,000, 1987, and digital data, 1:100,000, 1981–89;
Universal Transverse Mercator Projection, Zone 11. Shaded relief base from 1:250,000-scale Digital Elevation Model;
simulated sun illumination from northwest at 30 degrees above horizon

EXPLANATION

Water-level altitude, in feet above **NAVD 88**

☐ 1,760 to 1,780	▨ 1,840 to 1,860	☐ 2,220 to 2,240
☐ 1,780 to 1,800	▨ 2,080 to 2,100	☐ 2,240 to 2,260
☐ 1,800 to 1,820	▨ 2,180 to 2,200	☐ 2,260 to 2,880
☐ 1,820 to 1,840		

☐ Morongo
groundwater
subbasins

☐ Study area
boundary

⟍ Simulated horizontal-
flow barriers

–·–· USMC Air Ground
Combat Center
boundary

Figure 26. Simulated 2007 hydraulic-head contours for (*A*) model layer 1, (*B*) model layer 2, and (*C*) model layer 3 of the regional groundwater-flow model of the Twentynine Palms area, California.

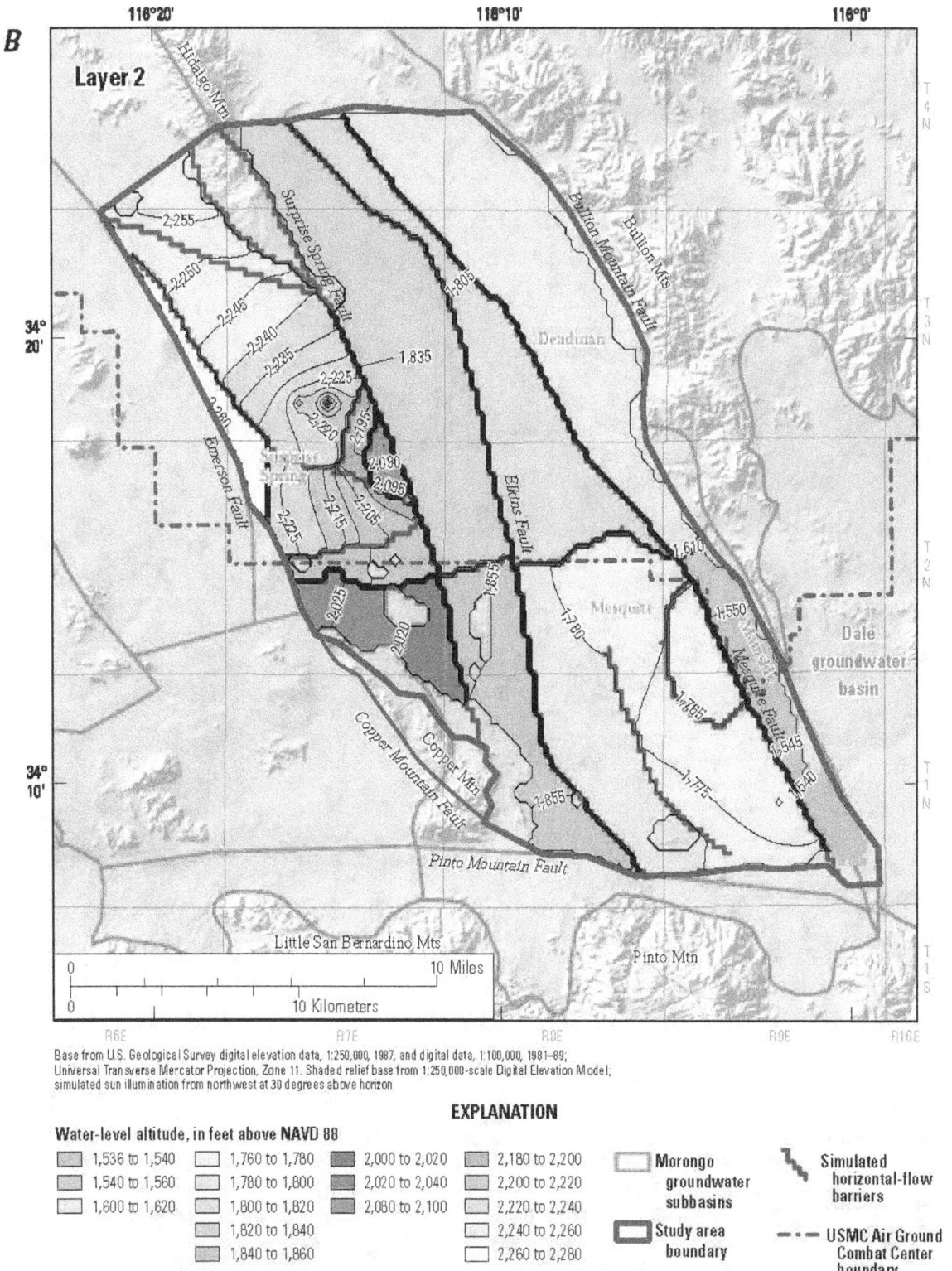

Base from U.S. Geological Survey digital elevation data, 1:250,000, 1987, and digital data, 1:100,000, 1981–89;
Universal Transverse Mercator Projection, Zone 11. Shaded relief base from 1:250,000-scale Digital Elevation Model;
simulated sun illumination from northwest at 30 degrees above horizon

EXPLANATION

Water-level altitude, in feet above NAVD 88

1,536 to 1,540	1,760 to 1,780	2,000 to 2,020	2,180 to 2,200
1,540 to 1,560	1,780 to 1,800	2,020 to 2,040	2,200 to 2,220
1,600 to 1,620	1,800 to 1,820	2,080 to 2,100	2,220 to 2,240
	1,820 to 1,840		2,240 to 2,260
	1,840 to 1,860		2,260 to 2,280

Morongo groundwater subbasins

Study area boundary

Simulated horizontal-flow barriers

USMC Air Ground Combat Center boundary

Figure 26.   Continued.

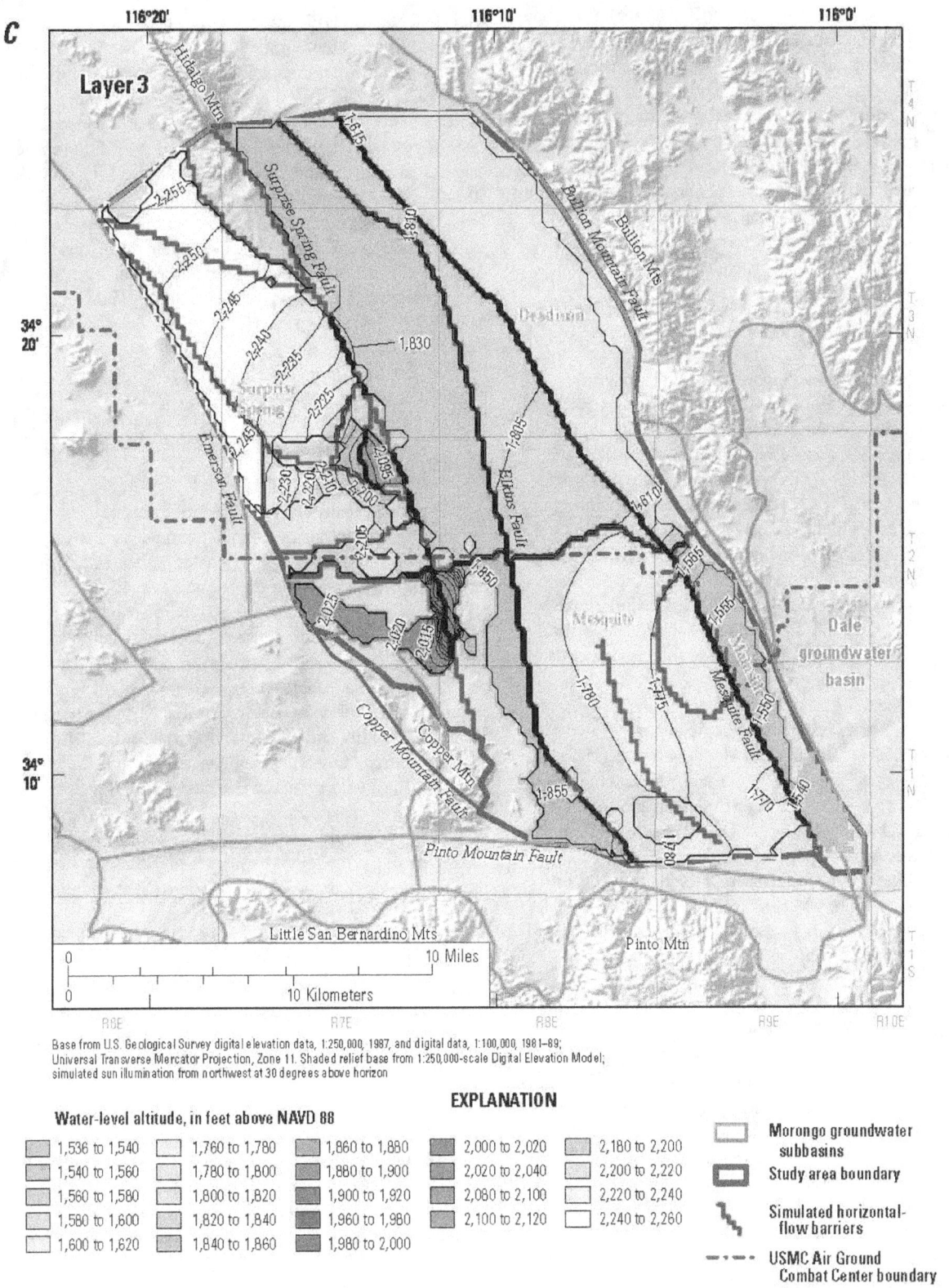

Base from U.S. Geological Survey digital elevation data, 1:250,000, 1987, and digital data, 1:100,000, 1981–89;
Universal Transverse Mercator Projection, Zone 11. Shaded relief base from 1:250,000-scale Digital Elevation Model;
simulated sun illumination from northwest at 30 degrees above horizon

EXPLANATION

Water-level altitude, in feet above NAVD 88

1,536 to 1,540	1,760 to 1,780	1,860 to 1,880	2,000 to 2,020	2,180 to 2,200
1,540 to 1,560	1,780 to 1,800	1,880 to 1,900	2,020 to 2,040	2,200 to 2,220
1,560 to 1,580	1,800 to 1,820	1,900 to 1,920	2,080 to 2,100	2,220 to 2,240
1,580 to 1,600	1,820 to 1,840	1,960 to 1,980	2,100 to 2,120	2,240 to 2,260
1,600 to 1,620	1,840 to 1,860	1,980 to 2,000		

☐ Morongo groundwater subbasins

☐ Study area boundary

Simulated horizontal-flow barriers

–·–·– USMC Air Ground Combat Center boundary

Figure 26. Continued.

about 139,400 acre-ft from the Surprise Spring subbasin, about 2,380 acre-ft from the Deadman subbasin, and about 3,670 acre-ft from the Mesquite subbasin. Analysis of the simulated groundwater-flow budgets indicates that almost all of the groundwater pumpage in the Surprise Spring subbasin originates as groundwater storage (fig. 27). Groundwater pumpage and associated depletion of groundwater storage (from the simulated change of storage) reached a maximum rate of about 4,200 acre-ft/yr in 2000. The simulated decline in hydraulic head in the Surprise Spring is the result of this depletion in groundwater storage (fig. 24B; appendix A, figs. A1–A5).

Simulated Interzonal Flows

The calibrated regional model was used to quantify groundwater flow between subbasins and hydrogeologic zones in the active model domain and between layers and hydrogeologic zones in the Surprise Spring subbasin. The USGS computer program ZONEBU–GET (Harbaugh, 1990) was used to calculate interzonal flow between each pair of neighboring hydrogeologic zones in the regional groundwater-flow model. The active model domain was divided into 15 zones by faults and (or) other groundwater barriers (fig. 28A). The Surprise Spring subbasin consists of zones 1–7, The Deadman subbasin consists of zones 9–11, the Mesquite subbasin consists of zones 8, 12, 13, and 15, and the Mainside subbasin consists of zone 14.

The calibrated model was used to determine the net groundwater flow between hydrogeologic zones for 2000 conditions. The simulated net groundwater flow between Surprise Spring and Mesquite subbasins was 0 acre-ft/yr and the simulated net groundwater flow between Deadman and Mesquite subbasins was 11 acre-ft/yr (fig. 28A). About 42 acre-ft/yr of net groundwater occured between the Surprise Spring Subbasin and the Deadman subbasin across the Surprise Spring Fault (fig. 28A). Groundwater flow occured across the Copper Mountain and Pinto Mountain Faults into the zones 8, 12, and 13 of the Mesquite subbasin and then discharged into zone 15 of the subbasin, where almost all of the flow is lost to evapotranspiration. The Mainside subbasin received only 5 acre-ft/yr of groundwater flow from the Mesquite subbasin and 9 acre-ft/yr from the Deadman subbasin (fig. 28A). The limited flow from the Mesquite subbasin into the Mainside subbasin reflects the low permeability (low hydraulic-characteristic value) of the Mesquite Fault (table 5).

Groundwater pumping in zones 2, 4, and 5 of the Surprise Spring subbasin has induced relatively high rates of groundwater flow from adjacent zones (fig. 28B). This can be of particular importance if the pumping is inducing the flow of groundwater from a zone that contains poor water quality. For example, high dissolved solids concentrations are present in zone 6 and high arsenic concentrations are present in zone 3 (figs. 14, 28). In addition, poor water quality is associated with the lower aquifer (layer 3) (Londquist and Martin, 1991). To better understand the movement of this poor-quality

water, ZONEBUDGET was used to calculate interzonal and interlayer flows in the Surprise Spring subbasin (fig. 28B). Groundwater flows horizontally from outer zones of the subbasin to the center of the pumping depression (zones 2, 4, and 5) in each layer. Pumping in zones 4 and 5 has induced high rates of groundwater flow from layer 2 of zone 3. Pumpage also has induced groundwater flow from zone 6 into zone 2. Because of the known poor water quality in zones 3 and 6, groundwater moving from these zones would likely degrade water quality in the receiving zones. Groundwater pumpage in zones 2, 4, and 5 has caused upward flow from layer 3—which also is reported to contain poor water quality (Londquist and Martin, 1991)—into layer 2. Temporal variations of horizontal flow rates from zone 6 to 2 in layer 3 and vertical flow rates from layer 3 to 2 for zones 2 and 5 are relatively small, but the trends have increased since the 1980s (fig. 29).

Model Sensitivity

Sensitivity analysis is a procedure that evaluates model sensitivity to variations in the input parameters. For this study, an automated sensitivity analysis, as described by Hill and others (2000), was not completed, because even a small change in some parameters, such as the hydraulic-characteristic value of certain fault segments, caused the model to fail to converge. To avoid numerical convergence problems, the sensitivity analysis was done using a series of forward simulations. A subgroup of model parameters was selected on the basis of numerical stability and hydraulic importance (table 7). For each simulation, only one parameter was perturbed upward and downward within a specific range. The resulting change in the Root Mean Square Error (RMSE) between the simulated hydraulic heads and the measured water levels reflects the sensitivity of the model to the perturbed parameter. In general, the larger the maximum absolute change of the parameter, the larger the parameter sensitivity.

For the regional model, the sensitivity of 36 parameters was tested, including hydraulic properties of three model layers, hydraulic-characteristic values of most major faults, and recharge from groundwater underflow (table 7). Each calibrated value of each parameter was multiplied by 0.2 and by 5 to test the sensitivity of the model to systematic changes in the parameter, for the simulation period 1953–2002 . A total of 1,622 water-level measurements from 34 wells in four groundwater subbasins were used to calculate the associated RMSE values. The sensitivity analysis was completed in 2002 using a preliminary version of the model and pre-2002 water-level measurements. Recent updates (2008) included modifying localized parameters and extending the model to 2007 conditions. The sensitivity analysis was not repeated using the updated model. Considering that changes in localized parameters are small compared with the parameters in the preliminary model, and that changes are focused in a specific time period and (or) area, their influence on the RMSE of the entire model is considered to be limited.

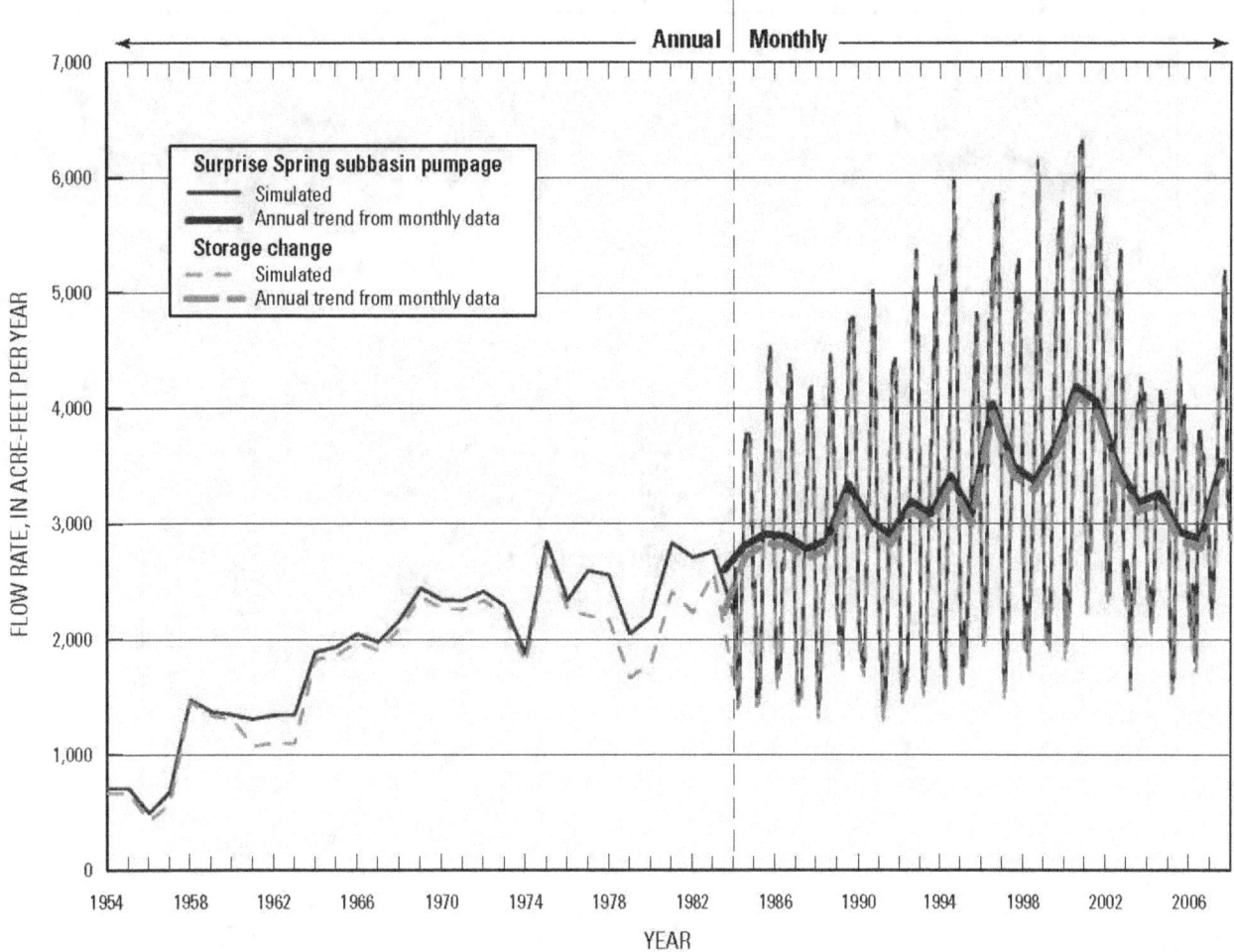

Figure 27.   Simulated rate in aquifer-storage change, as compared with the U.S. Marine Corps Air Ground Combat Center (MCAGCC) pumpage, for the Surprise Spring subbasin of the regional groundwater-flow model of the Twentynine Palms area, California, 1954–2007.

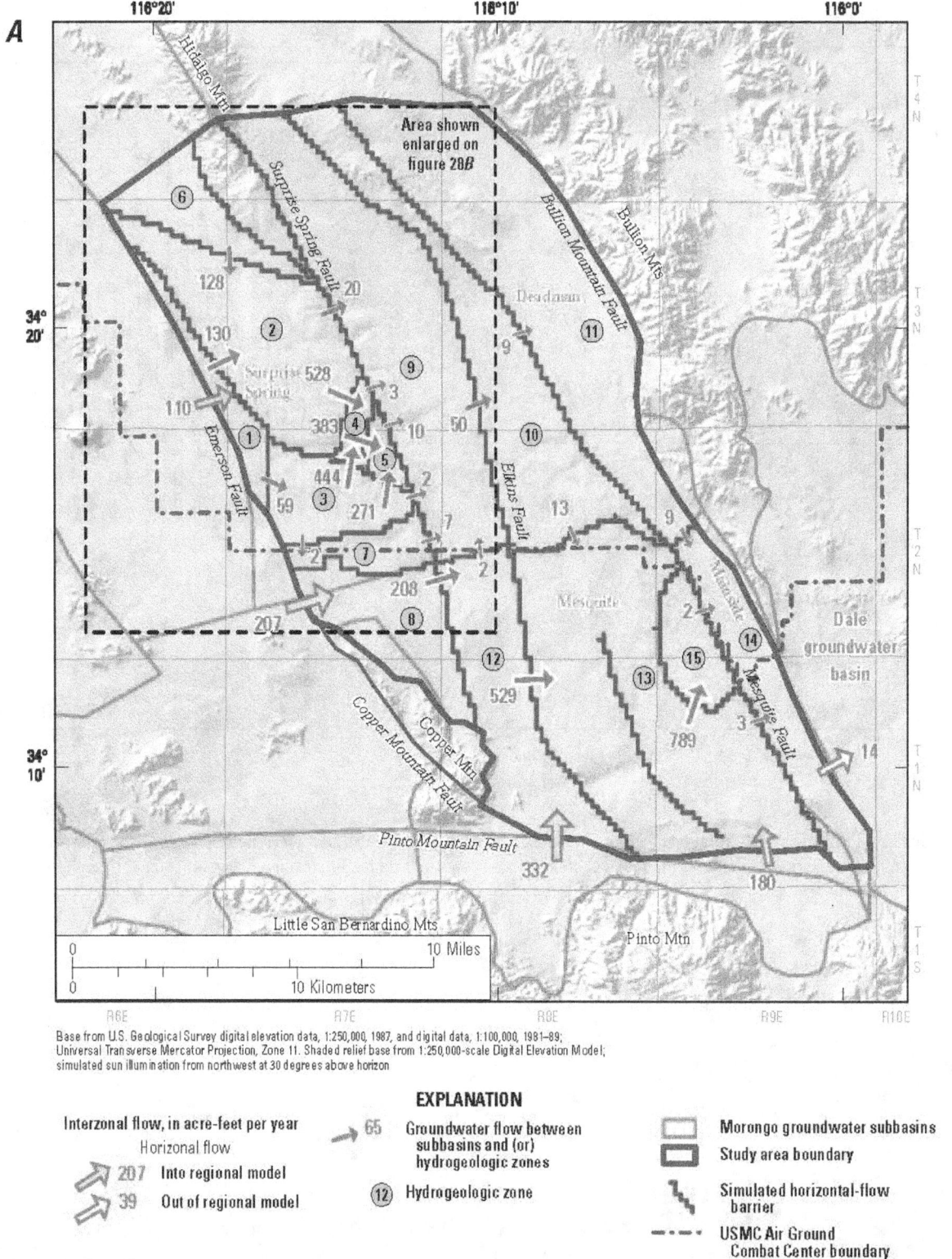

Base from U.S. Geological Survey digital elevation data, 1:250,000, 1987, and digital data, 1:100,000, 1981–89;
Universal Transverse Mercator Projection, Zone 11. Shaded relief base from 1:250,000-scale Digital Elevation Model;
simulated sun illumination from northwest at 30 degrees above horizon

EXPLANATION

Interzonal flow, in acre-feet per year
   Horizonal flow
   207 Into regional model
   39 Out of regional model

→ 65 Groundwater flow between
   subbasins and (or)
   hydrogeologic zones

⑫ Hydrogeologic zone

☐ Morongo groundwater subbasins

▭ Study area boundary

⌐ Simulated horizontal-flow
   barrier

–·–·–· USMC Air Ground
   Combat Center boundary

Figure 28. Simulated interzonal flows between (A) subbasins and hydrogeologic zones in the active model domain and
(B) between layers and hydrogeologic zones in the Surprise Spring subbasin for 2000 conditions of the regional groundwater-
flow model of the Twentynine Palms area, California.

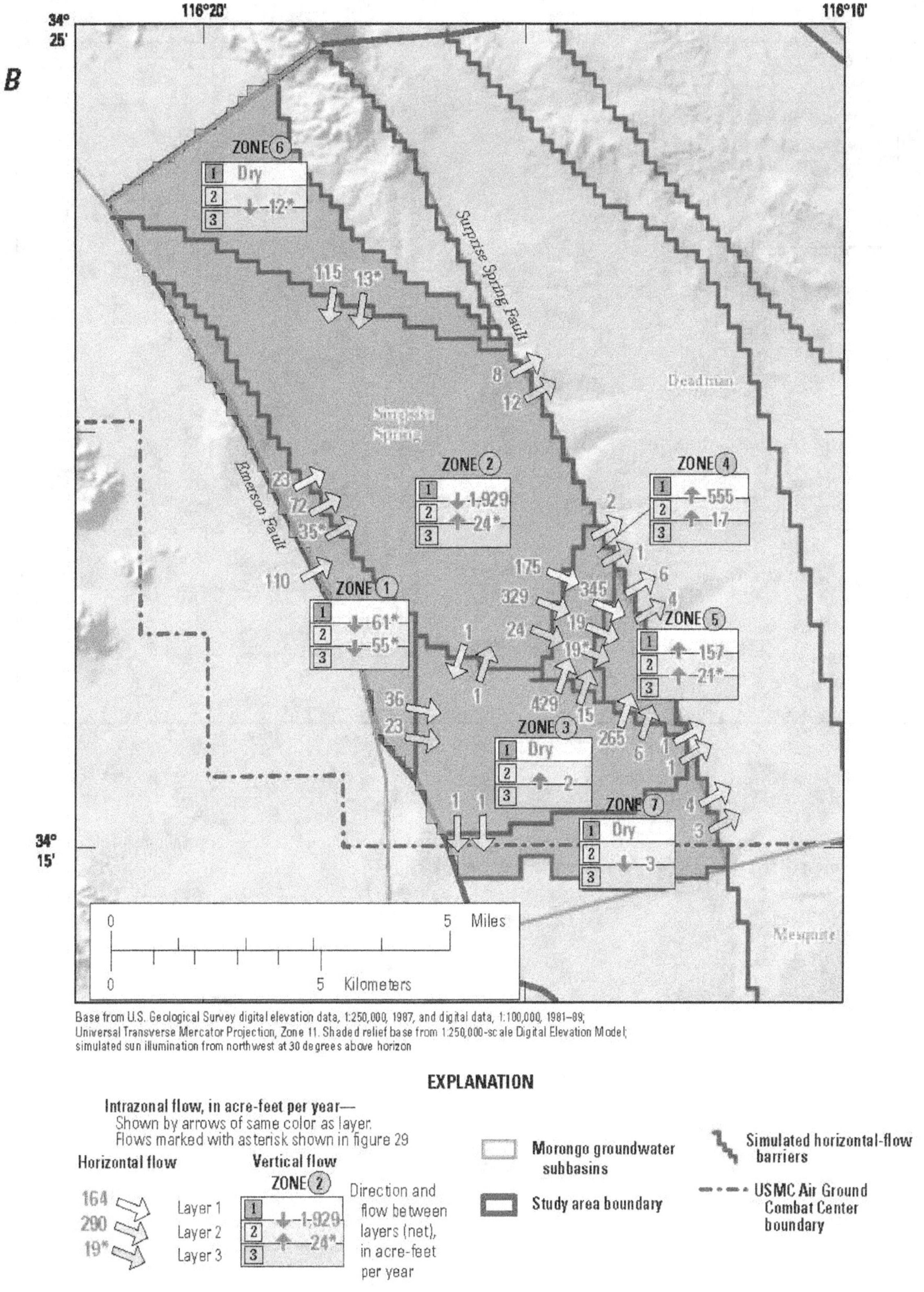

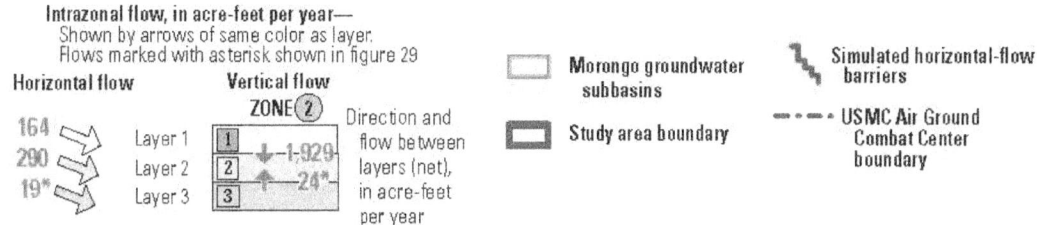

Figure 28.   Continued.

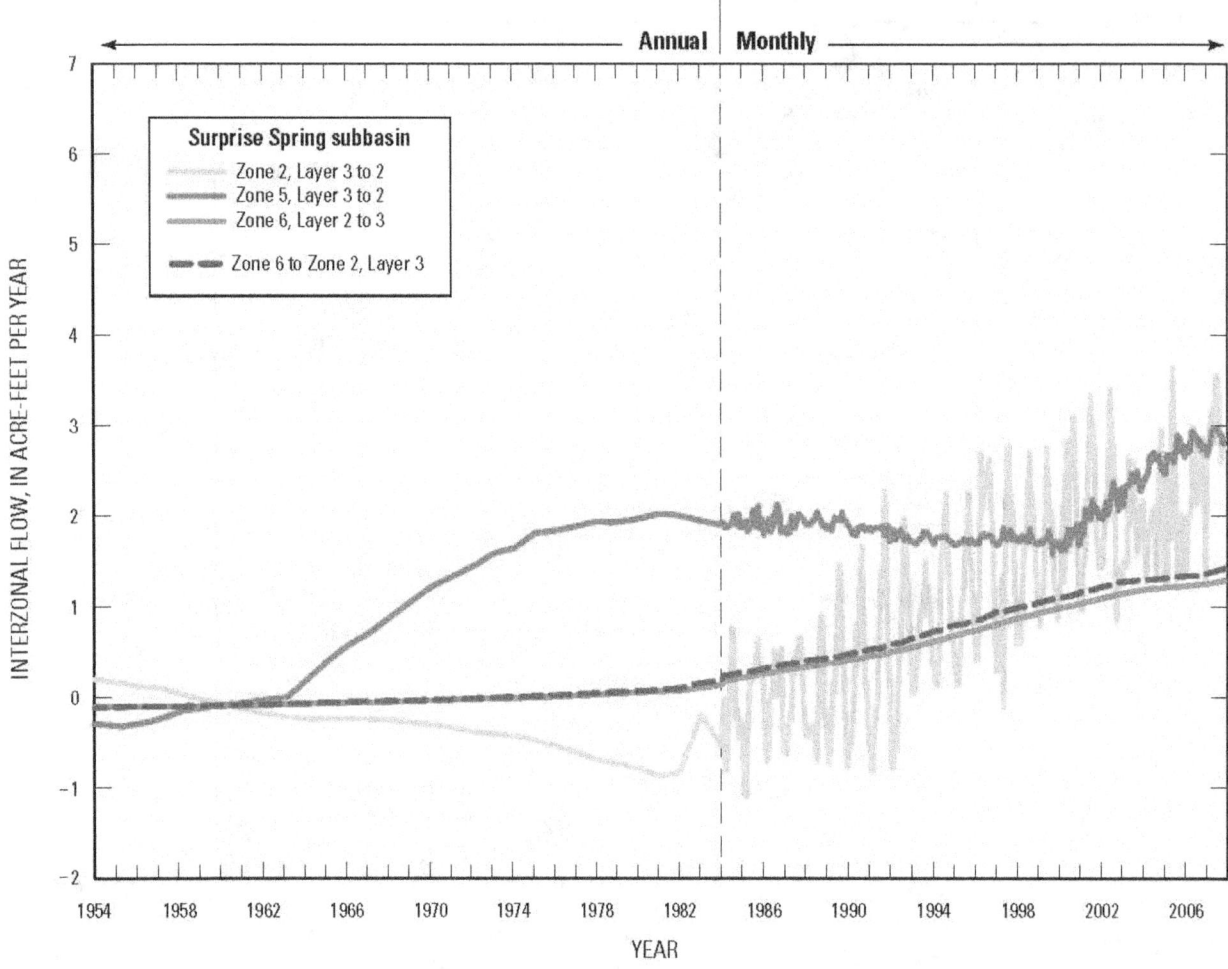

Figure 29. Simulated interzonal flow between selected hydrogeologic zones and layers in the Surprise Spring subbasin of the regional groundwater-flow model of the Twentynine Palms areas, California, 1953–2007.

Table 7. Sensitivity of selected model parameters for the regional groundwater-flow model of the Twentynine Palms area, California.

[**Abbreviations:** NC, the sensitivity run was not complete because of numerical convergence failure; RMSE, root mean squared error, in feet; SS, Surprise Spring subbasin; SSF, Surprise Spring Fault; MQF, Mesquite Fault; ELKF, Elkin Fault; HydChar, Hydraulic Characteristic]

Parameter name	Change in RMSE from calibrated estimate		Maximum absolute change in RMSE
	0.2X	5X	
Specific storage, layer 1	−2.86E−04	−5.92E−04	5.92E−04
Specific storage, layer 2	−1.50E−03	−2.28E−03	2.28E−03
Specific storage, layer 3	NC	7.97E−03	7.97E−03
Specific yield, layer 1	NC	7.54E+00	7.54E+00
Specific yield, layer 2	2.79E+00	1.54E+00	2.79E+00
Specific yield, layer 3	−6.95E−04	7.31E−05	6.95E−04
Horizontal hydraulic conductivity, layer 1	−2.42E−01	NC	2.42E−01
Horizontal hydraulic conductivity, layer 2	NC	1.50E+00	1.50E+00
Horizontal hydraulic conductivity, layer 3	−1.35E−01	1.19E+00	1.19E+00
Vertical hydraulic conductivity, layer 1	−9.39E−03	3.21E−03	9.39E−03
Vertical hydraulic conductivity, layer 2	3.85E−01	1.13E−01	3.85E−01
Vertical hydraulic conductivity, layer 3	8.58E−01	−5.06E−02	8.58E−01
Western Inflow 1	−9.44E−03	6.36E−02	6.36E−02
Western Inflow 2	−5.90E−03	3.14E−02	3.14E−02
Western Inflow 3	−2.48E−02	2.46E−01	2.46E−01
Western Inflow 4	−6.39E−03	3.50E−02	3.50E−02
Western Inflow 5	−4.40E−03	8.84E−02	8.84E−02
Southern Inflow 1	−1.94E−03	1.06E−02	1.06E−02
Southern Inflow 2	−1.33E−03	7.15E−02	7.15E−02
HydChr of Fault Segment k (inside SSB), layer 2	−6.39E−04	−7.79E−04	7.79E−04
HydChr of Fault Segment k (inside SSB), layer 3	2.18E−04	−5.26E−04	5.26E−04
HydChr of Fault Segment r (MQF), layer 2	−7.89E−05	−3.82E−03	3.82E−03
HydChr of Fault Segment r (MQF), layer 3	2.44E−02	−8.88E−02	8.88E−02
HydChr of Fault Segment e (upper SSF), layer 2	−2.40E−03	4.01E−02	4.01E−02
HydChr of Fault Segment e (upper SSF), layer 3	1.04E−03	2.72E−02	2.72E−02
HydChr of Fault Segment o (lower ELKF), layer 2	−4.37E−04	6.86E−04	6.86E−04
HydChr of Fault Segment o (lower ELKF), layer 3	−9.25E−03	3.59E−02	3.59E−02
HydChr of Fault Segment g (inside SSB), layer 2	−3.47E−04	−1.87E−03	1.87E−03
HydChr of Fault Segment g (inside SSB), layer 3	−7.09E−04	−4.26E−04	7.09E−04
HydChr of Fault Segment a (inside SSB), layer 1	4.71E−03	−1.59E−02	1.59E−02
HydChr of Fault Segment a (inside SSB), layer 2	3.33E−02	−3.01E−02	3.33E−02
HydChr of Fault Segment a (inside SSB), layer 3	3.49E−03	−1.61E−03	3.49E−03
HydChr of Fault Segment m (lower SSF), layer 2	1.67E−01	7.83E−01	7.83E−01
HydChr of Fault Segment m (lower SSF), layer 3	1.04E−03	2.72E−02	2.72E−02
HydChr of Fault Segment n (upper ELKF), layer 2	1.99E−02	2.28E−01	2.28E−01
HydChr of Fault Segment n (upper ELKF), layer 3	−1.78E−02	1.02E−02	1.78E−02

Specific storage, specific yield, and hydraulic conductivity are areal parameters. The multiplication factors (0.2 and 5) were applied to all zones containing the tested areal parameter. For the areal parameters, the model was most sensitive to the specific yields of model layers 1 and 2. The model also was sensitive to changes in hydraulic conductivity. In general, horizontal and vertical hydraulic conductivity sensitivities of layer 2 and 3 are higher than that of layer 1. Simulated hydraulic heads in layer 1 were less sensitive to changes in hydraulic conductivity because layer 1 has a higher specific yield value than the other layers (table 3).

Fault hydraulic-characteristic values and groundwater underflow are localized parameters. The hydraulic-characteristic value of fault segment **m**, layer 2 (the lower part of the Surprise Spring Fault) is the most sensitive one of all fault segments being simulated (fig. 19, table 7). This fault segment controls underflow from the Joshua Tree groundwater basin and recharge to the broad area in between the Surprise Spring and the Mesquite Faults in the Mesquite subbasin (fig. 28A). The model also was sensitive to the hydraulic-characteristic value of fault segment **n**, layer 2 (upper part of Elkin Fault) (fig. 19, table 7). This fault segment controls groundwater flow through the Deadman subbasin (fig. 28A). Of the underflows, the model was most sensitive to the underflow across the south-central section of the Emerson Fault (Western Inflow 3, table 7). This underflow has a relatively large effect on the model RMSE because it is closer to a larger number of observations than other underflows and to the more actively pumped Surprise Spring subbasin.

In general, model sensitivity is influenced by parameter definition, location of observation wells, and boundary conditions. Areally extensive parameters can appear to be more sensitive than more localized parameters because more measurements can be affected directly. Sensitivity of localized parameters tends to increase as the number of observations near that parameter increases.

Model Use and Limitations

As designed and calibrated, the regional groundwater-flow model of the MCAGCC is best used for analyzing regional issues of water use and supply. The model is particularly useful for estimating changes in regional groundwater levels and flows in response to groundwater extraction and artificial recharge. The model also can be used to define boundary conditions for local models of smaller domains contained within the regional model. Direct use for site-specific, short period interpretation is not recommended because the model was developed to focus on problems of regional and long-term scales.

A model only is as good as the data that were used to develop it. The accuracy and reliability of model prediction is related to the quality and distribution of available data. For areas with long-term groundwater development, such as the Surprise Spring subbasin, sufficient data are available to characterize aquifers and to calibrate for water-level variations in time and space. For areas that have limited data, the uncertainty of model predictions is increased. For example, model uncertainty for the northeastern part of Deadman subbasin is generally higher than for other simulated areas because almost no data, or only sporadic data, are available for this part of the model domain. Similarly, model uncertainty for the deep aquifer of the Mesquite subbasin is large because limited data from a single deep production well in the Mesquite subbasin (1N/9E–21H1, table 1) was used to estimate constant hydraulic properties for each layer of the subbasin.

Except in the Surprise Spring subbasin, there has been limited groundwater development in the model domain. Therefore, few data constrain the calibration of the transient model in the Deadman, Mesquite, and Mainside subbasins. The hydraulic properties (hydraulic conductivity, specific yield, and specific storage) need to be updated and refined for these subbasins as data become available. As shown in this report, faults can have a significant barrier effect on the flow of groundwater in the study area. Therefore, in order to accurately simulate groundwater flow, the location and hydraulic properties of the faults must be well understood. However, the locations and geometries of faults within the model domain are uncertain because limited data are available for most of the study area. As more information becomes available, the locations of faults may be revised and additional faults may need to be included in the model.

Limited depth-dependent water-level data is a common problem for many groundwater basins. Existing multiple-well monitoring sites in the Surprise Spring, Deadman, and Mainside subbasins (fig. 8) provided valuable depth-dependent water-level and water-quality information. To better understand vertical flows in the region, more depth-dependent water-level data are needed, especially in the Surprise Spring and Mesquite subbasins.

Seismic events have affected water levels measured in several monitoring wells in the lower aquifer in the study area (Roeloffs and others, 2001) (appendix A, fig. A5, well 3N/7E–32D3). The regional model developed for this study does not simulate the effects of seismic events on hydraulic heads.

Irrigation return flows are a potential source of recharge to the study area in the Mainside and Mesquite subbasins. The regional model did not include this potential source of recharge because available data indicate that return flows have not affected water levels in the Mainside subbasin and have only recently (2000) affected water levels near the MCAGCC golf course in the Mesquite subbasin. Future model runs may need to include irrigation return flows as a source of recharge if water-level and water-quality data indicate that the flows have migrated to the water table.

A numerical model only is an approximation of the actual system that is based on average and estimated conditions. The accuracy and the reliability of the model is dependent on the accuracy of the data used to build the model and the adequacy of the model to simulate the actual system. Despite limitations of the data and the numerical method, no other approach is better than a groundwater model to integrate a wide variety of data from multiple sources and to develop concepts of a largely unseen system. Groundwater modeling is an iterative process with data and simulation complementing each other. As more data are collected, the regional model developed for this study could be updated and recalibrated to improve the understanding of the aquifer system.

Assessment of Water-Management Strategies

The MCAGCC is considering various water-management strategies to manage their limited water resources in the Twentynine Palms area. A variety of strategies have been considered, including conservation, more effective use of potable water, and development of new water sources. Proper development and use of potable and nonpotable water from various sources may allow the MCAGCC to meet growing demand and Federal and State water-quality standards, while also limiting overdraft in the Surprise Spring subbasin. The calibrated regional groundwater model was used to simulate the hydrologic effects of the strategies being considered by the MCAGCC.

Water-Management Challenges

The MCAGCC is located in a remote, desert area, with groundwater as the only source of supply. Historically, the MCAGCC has relied on groundwater pumped from the upper and middle aquifers in the Surprise Spring subbasin to serve all of its needs. Groundwater from the lower aquifer of this and nearby subbasins (Deadman, Mainside, and Mesquite) contains high concentrations of fluoride, arsenic, and (or) dissolved solids (Riley and Worts, 1953 [2001]; Londquist and Martin, 1991), making it unsuitable for potable uses without treatment.

From 1953 to 2007, the MCAGCC pumped about 139,400 acre-ft of potable groundwater from the Surprise Spring subbasin. Almost all of this pumped groundwater was removed from aquifer storage, resulting in water-levels declining as much as 190 ft in the Surprise Spring subbasin near the Surprise Spring. The groundwater pumping and resulting water-level declines have decreased the saturated thickness of the productive upper aquifer by almost 50 percent since groundwater pumping started in the inner hydrogeologic zone near the ancestral Surprise Spring.

The MCAGCC has initiated mitigation measures to reduce water-level declines in the Surprise Spring subbasin. Measures taken included re-allocating groundwater pumping to the middle and outer hydrogeologic zones of the Surprise Spring subbasin (figs. 8, 10), conserving water, utilizing treated wastewater for irrigation instead of potable water

from the Surprise Spring subbasin, and pumping nonpotable groundwater in the Mainside subbasin to supplement treated wastewater during the summer months as a source of golf-course irrigation (table 8). These efforts have helped to reduce groundwater overdraft in the Surprise Spring subbasin, but groundwater pumpage from the subbasin (about 3,300 acre-ft/yr) continues to greatly exceed estimates of natural recharge to the basin (110 acre-ft/yr) (fig. 10, table 6). Recent investigations of water usage by the MCAGCC (Robert Lehman, Chief of Engineering, U.S. Marine Corps Air Ground Combat Center, written commun., 2008) indicate that about one-third of the potable water pumped from the Surprise Spring subbasin currently is being used for nonpotable use (table 9).

Results from the calibrated model show that as the saturated thickness of the upper aquifer of the Surprise Spring subbasin decreases, the specific capacity of the wells in the subbasin will decrease, and the wells will take an increasing percentage of water from the lower aquifer, which contains poor quality water (high concentrations of total dissolved solids, fluoride, and arsenic). The 2006 reduction in the U.S. Environmental Protection Agency (USEPA) Maximum Contaminant Level (MCL) for arsenic from 50 to10 µg/L (U.S. Environmental Protection Agency, 2010a) has reduced the available groundwater resources in the study area that are suitable for potable use without treatment.

Water demand in the Twentynine Palms area continues to increase as the population increases. The MCAGCC projects that population growth of the permanent and transient military personnel and civilian workforce could be as high as 20 percent over the next 10 years (Robert Lehman, Chief of Engineering, U.S. Marine Corps Air Ground Combat Center, written commun., 2008). The city of Twentynine Palms activated a major production well in the Mesquite subbasin in 2003, and a second production well is scheduled to be on line in the Mesquite subbasin in the near future. For the purposes of this study, population growth and associated water demand of the city of Twentynine Palms are assumed to be 2 percent annually.

On January 24, 2007, President George W. Bush signed Executive Order 13423, "Strengthening Federal Environmental, Energy, and Transportation Management." One goal of the executive order is to reduce water consumption of Federal agencies relative to the baseline of an agency's water consumption in Federal fiscal year 2007 (FY2007; October 1, 2006–September 30, 2007) by 2 percent annually. During FY2007, the Surprise Spring subbasin pumpage was about 3,550 acre-ft, which is the baseline for the MCAGCC pumpage from the Surprise Spring subbasin.

Water-Management Scenarios

The calibrated regional groundwater-flow model was used to help evaluate four water-management scenarios being considered by the MCAGCC to meet the projected water demand at the base over the next 10 years (2008–2017).

Table 8. Reported monthly volume of U.S. Marine Corps Air Ground Combat Center (MCAGCC) golf course water demand, evaporation at Ocotillo pond, wastewater treatment plant water, and golf course well pumpage required to supplement golf course irrigation in the Twentynine Palms area, California.

[Values are estimated and reported by MCAGCC by assuming maximum evaporation and well production rate from April to October. **Abbreviation:** WWTP, Wastewater treatment plant]

Month (end)	Golf course demand (million gallons)	Ocotillo pond evaporation (million gallons)	WWTP supply qualified for irrigation (million gallons)	Golf course well pumpage to supplement WWTP supply (million gallons)
January	9.8	0.1	9.8	0.0
February	13.3	.1	13.4	.0
March	21.0	.2	21.2	.0
April	27.5	.2	8.6	19.1
May	34.8	.3	8.1	27.0
June	38.4	.3	10.6	28.1
July	38.7	.3	10.5	28.5
August	34.8	.3	8.1	27.0
September	27.5	.3	.7	27.0
October	20.0	.2	1.1	19.1
November	11.8	.1	11.9	.0
December	8.8	.1	8.9	.0
Total in millions of gallons	286.3	2.5	112.9	175.8
Total in acre-feet	878.6	7.5	346.6	539.5

Table 9. Estimated U.S. Marine Corps Air Ground Combat Center nonpotable water usage in the Twentynine Palms area, California.

[Estimates are from Robert Lehman, Chief of Engineering, U.S. Marine Corps Air Ground Combat Center, written commun., 2008]

Usage	Gallons per day	Million gallons per year	Acre-feet per year
Golf course irrigation supplement	481,644	175.80	540
Mainside landscaping	402,934	147.07	451
Mainside vehicle wash rack	13,000	3.64	11
Construction activities	100,285	28.08	86
Cooling towers/evaporative coolers	88,000	18.83	58
Aircraft wash rack	6,000	1.5	5
Total	1,091,863	374.92	1,151

Specifically, the model was used to estimate the effects of proposed water management scenarios on water levels and aquifer conditions in the Surprise Spring, Deadman, Mesquite, and Mainside subbasins of the Morongo groundwater basin. The proposed scenarios involve modifying the quantity and distribution of the MCAGCC pumpage from the Surprise Spring, Deadman, and Mainside subbasins. A primary goal of the water-management scenarios is to replace pumpage of potable water in the Surprise Spring subbasin with pumpage of nonpotable groundwater in the Deadman and the Mainside subbasins. The water-management scenarios evaluated in this report only represent a small number of the water-mangement strategies being considered by the MCAGCC to help manage their limited water resources.

Description of Model Scenarios

Groundwater recharge for all of the model scenarios was assumed to be the same as that calibrated for the steady-state model (table 6). Actual recharge in future years may be less than the steady-state value because of increased groundwater usage from subbasins upgradient of the Morongo groundwater basin and possible long-term drought conditions. Groundwater recharge from irrigation-return flow was assumed not to reach the water table during the period simulated in the scenarios; therefore, irrigation-return flow was not simulated in the scenarios. Natural discharge (spring discharge, evapotranspiration, and groundwater underflow) is head-dependent and was simulated by the groundwater-flow model. Discharge through pumping was projected for each scenario at each simulated year (2008 -2017) starting with 2007 reported pumpage.

Scenario 1

Scenario 1 assumes a 20-percent increase in population at the MCAGCC from 2008 to 2017, which approximately translates into an annual increase in the MCAGCC water demand and pumpage of slightly less than 2 percent. The scenario also assumes a 20-percent increase in Twentynine Palms pumpage from the Mesquite subbasin. Pumpage from the Mainside subbasin for irrigating the MCAGCC golf course was assumed to be constant at the 2007 rate (540 acre-ft/yr) for 2008–2017. Projected pumpage in the Surprise Spring, Mainside, and Mesquite subbasins reached a maximum at the end of the simulation (2017) of about 4,260 acre-ft/yr, 540 acre-ft/yr, and 920 acre-ft/yr, respectively.

Projected annual groundwater pumpage in the Surprise Spring subbasin was distributed to each of the active MCAGCC production wells in the subbasin based on the percentage of average monthly pumpage for each well over the period 2000 to 2004 (fig. 30, table 10). The monthly reported production rate of the golf course well (table 8) was assumed constant for all years of the scenario. All of the pumpage from the golf course well occurs between April and October (table 8). The 2007 monthly production for the City of Twentynine Palms well was assumed to be representative of the pumping

pattern for the scenario. Table 11 shows the monthly pumpage distribution by well for year 1 (2008) of water-management scenario 1. The total projected cumulative pumpage for Scenario 1 is about 53,670 acre-ft, with more than 70 percent of the pumpage from the Surprise Spring subbasin (fig. 31). The location of Scenario 1 pumping wells is shown on figure 32A (presented lated in the report).

Scenario 2

Scenario 2 assumes the same 20-percent increase in population as simulated in Scenario 1; however, Scenario 2 assumes that the MCAGCC will institute a water-conservation plan that will conserve 2 percent of its potable water supply from the Surprise Spring subbasin annually, as mandated by Executive Order 13423. The water conservation offsets the projected increase in the MCGACC potable water supply; therefore, Scenario 2 maintains 2007 MCAGCC pumpage from the Surprise Spring subbasin throughout the simulation period (2008–2017). Pumpage from the MCAGCC golf course well in the Mainside subbasin was assumed to be constant at 2007 rates throughout the scenario. Projected pumpage for the City of Twentynine Palms was estimated using the same water-conservation assumptions used for the MCAGCC pumpage projections. Projected pumpage in Scenario 2 from the Surprise Spring, Mainside, and Mesquite subbasins remains at the 2007 level to the end of the simulation (2017).

Similar to Scenario 1, the percentage of average monthly pumpage during 2000-2007 (table 10) was used to calculate the monthly distribution of pumpage for Scenario 2 for the MCAGCC supply wells in the Surprise Spring subbasin. The monthly distribution of pumpage for the golf course supply well in the Mainside subbasin, and the Twentynine Palms supply well in the Mesquite subbasin were the same as 2007. Table 12 shows the monthly pumpage distribution by well for Scenario 2. The total projected cumulative pumpage for Scenario 2 is about 49,420 acre-ft (fig. 31). Location of scenario 2 pumping wells is illustrated in figure 32B (presented lated in the report).

Scenario 3

Scenario 3 assumes the same water-conservation plan as Scenario 2, and adds measures to replace potable water pumped from the Surprise Spring subbasin for nonpotable uses with nonpotable water pumped from the Deadman and the Mainside subbasins. For Scenario 3, two nonpotable supply wells (CW1 and CW2) were simulated in the middle hydrogeologic zone of the Deadman subbasin and six nonpotable supply wells were simulated in the Mainside subbasin (MS1, MS2, MS3, MS4, MS5, and MS6). All of the simulated nonpotable supply wells were assumed to pump solely from the middle aquifer (model layer 2), which is saturated in most of the study area.

In 2007, about 610 acre-ft/yr of water was pumped from the Surprise Spring subbasin for nonpotable use (irrigating Mainside subbasin green areas, washing military equipment,

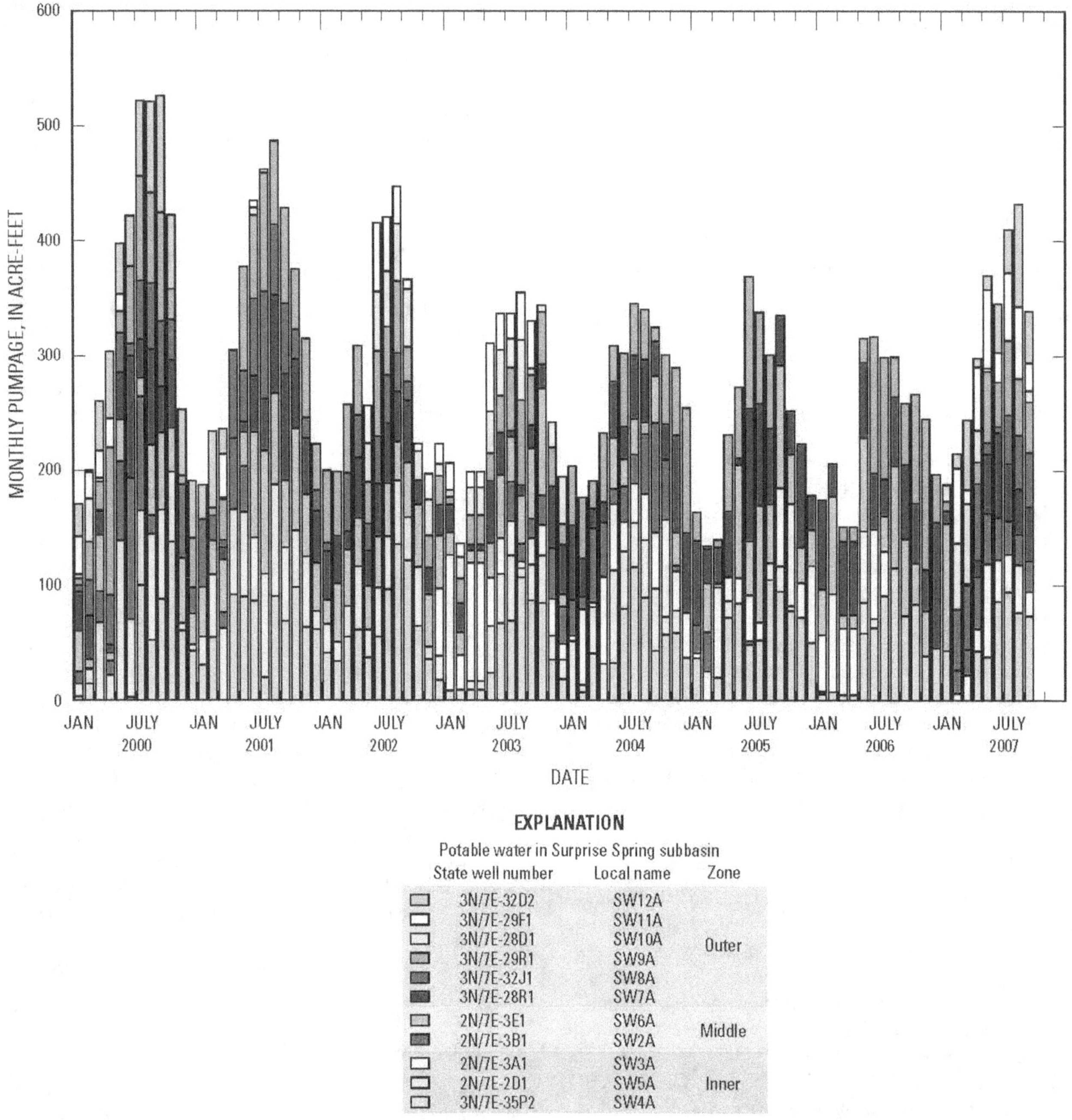

EXPLANATION

Potable water in Surprise Spring subbasin

State well number	Local name	Zone
3N/7E-32D2	SW12A	
3N/7E-29F1	SW11A	
3N/7E-28D1	SW10A	Outer
3N/7E-29R1	SW9A	
3N/7E-32J1	SW8A	
3N/7E-28R1	SW7A	
2N/7E-3E1	SW6A	Middle
2N/7E-3B1	SW2A	
2N/7E-3A1	SW3A	
2N/7E-2D1	SW5A	Inner
3N/7E-35P2	SW4A	

Figure 30. Measured pumpage of the U.S. Marine Corps Air Ground Combat Center (MCAGCC) supply wells in the Surprise Spring subbasin of the Twentynine Palms area, California, January 2000–September 2007.

Table 10.   Percentage of average monthly pumpage for U.S. Marine Corps Air Ground Combat Center supply wells in the Surprise Spring subbasin of the Twentynine Palms area, California, 2000–2004.

[All values are expressed as percentages except where noted; because of rounding, not all totals equal 100 percent]

Well name	Jan.	Feb.	Mar.	Apr.	May	June	July	Aug.	Sept.	Oct.	Nov.	Dec.
				Surprise Spring subbasin								
				Inner zone								
2N7E03A1–SW3A	13	10	9	13	6	3	3	2	4	7	5	10
3N7E35P2–SW4A	9	12	9	12	11	15	19	22	21	27	19	16
2N7E02D1–SW5A	12	10	18	14	22	13	14	14	13	13	9	5
				Middle zone								
2N7E03B1–SW2A	4	1	9	9	7	8	8	4	6	5	4	3
2N7E03E1–SW6A	12	11	14	4	11	11	11	11	14	17	14	12
				Outer zone								
3N7E28R1–SW7A	17	15	6	6	11	14	9	13	12	12	16	17
3N7E32J1–SW8A	7	6	5	12	6	6	9	7	7	4	6	4
3N7E29R1–SW9A	13	18	15	20	12	16	16	15	13	11	15	22
3N7E28D1–SW10A	1	6	2	2	4	5	4	5	3	2	4	5
3N7E29F1–SW11A	6	4	7	3	6	5	3	3	3	0	2	2
3N7E32D2–SW12A	6	7	6	4	3	2	3	4	5	4	4	5

Table 11. Monthly pumpage distribution by well for year 1 (2008) of water-management scenario 1 for the Twentynine Palms area, California.

[All values are in acre-feet]

Well Name	January	Febuary	March	April	May	June	July	August	September	October	November	December	Subtotal
Surpise Spring Subbasin													
Inner zone													
2N7E03A1–SW3A	24.1	22.2	23.2	40.0	23.0	10.9	13.4	10.1	12.9	18.2	13.0	20.6	
3N7E35P2–SW4A	16.4	27.2	22.8	36.4	41.6	53.1	79.7	96.1	70.5	71.2	48.2	32.5	
2N7E02D1–SW5A	22.2	20.9	44.2	42.5	81.0	47.2	58.8	61.7	46.0	35.2	23.1	11.1	1321.2
Middle zone													
2N7E03B1–SW2A	8.4	3.1	23.0	27.4	28.0	28.4	31.9	17.7	19.9	13.0	9.7	6.4	
2N7E03E1–SW6A	23.4	24.9	33.5	13.8	40.7	39.7	47.8	46.2	48.0	44.5	35.5	23.1	637.8
Outer zone													
3N7E28R1–SW7A	32.5	31.7	15.2	19.0	42.0	48.6	38.4	56.3	41.1	31.1	40.1	33.7	
3N7E32J1–SW8A	14.4	12.4	13.0	35.1	23.6	22.2	37.4	31.4	23.6	10.3	16.0	7.7	
3N7E29R1–SW9A	24.9	39.1	36.7	61.4	45.7	56.7	67.0	67.5	46.0	30.1	37.8	43.0	
3N7E28D1–SW10A	2.2	13.2	6.5	5.8	16.3	18.4	15.8	21.4	10.2	5.5	10.6	9.8	
3N7E29F1–SW11A	11.2	8.4	16.7	9.1	24.7	18.3	14.3	15.6	9.2	1.6	4.7	3.5	
3N7E32D2–SW12A	11.6	16.0	14.3	13.5	10.6	8.6	13.6	16.8	18.1	10.8	11.3	9.5	1660.0
Total													3619.1
Mainside Subbasin													
Golf course well	0.0	0.0	0.0	58.6	82.9	86.2	87.5	82.9	82.9	58.6	0.0	0.0	539.5
Mesquite Subbasin													
City of Twentynine Palms well	65.5	60.7	70.7	68.8	73.9	77.1	83.5	83.5	83.4	70.8	62.2	60.2	860.3

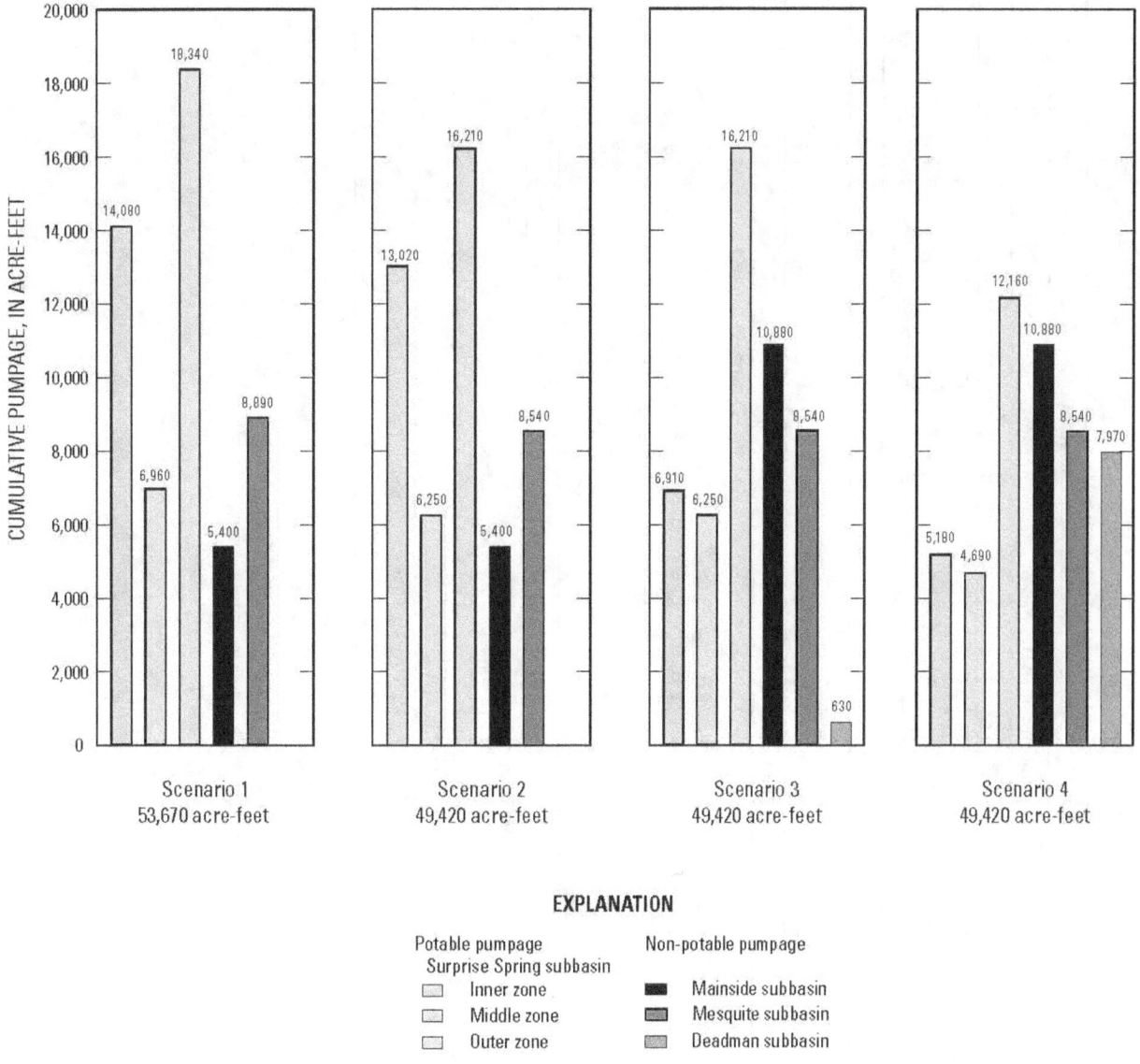

Figure 31. Total cumulative pumpage by subbasin for water-management scenarios 1-4 for the Twentynine Plams area, California.

Table 12. Monthly pumpage distribution by well for water-management scenario 2 for the Twentynine Palms area, California.

[All values are in acre-feet]

Well name	January	February	March	April	May	June	July	August	September	October	November	December	Subtotal
							Surprise Spring subbasin						
							Inner zone						
2N7E03A1–SW3A	23.8	21.8	22.8	39.4	22.3	10.2	12.7	9.3	12.3	17.7	12.5	20.3	
3N7E35P2–SW4A	16.1	26.8	22.4	35.9	41.0	52.5	78.9	95.4	69.9	70.7	47.7	32.1	
2N7E02D1–SW5A	21.9	20.5	43.8	41.9	80.3	46.6	58.1	60.9	45.4	34.8	22.6	10.7	1,301.9
							Middle zone						
2N7E03B1–SW2A	8.1	2.7	22.6	26.9	27.3	27.8	31.2	16.9	19.3	12.5	9.3	6.1	
2N7E03E1–SW6A	23.0	24.6	33.0	13.2	40.0	39.1	47.0	45.4	47.4	44.0	35.0	22.7	624.9
							Outer zone						
3N7E28R1–SW7A	32.1	31.3	14.8	18.4	41.3	47.9	37.7	55.5	40.5	30.6	39.7	33.3	
3N7E32I1–SW8A	14.1	12.0	12.5	34.6	23.0	21.6	36.6	30.6	22.9	9.8	15.5	7.3	
3N7E29R1–SW9A	24.6	38.7	36.3	60.8	45.0	56.0	66.2	66.7	45.4	29.7	37.4	42.7	
3N7E28D1–SW10A	1.9	12.8	6.1	5.3	15.7	17.8	15.1	20.6	9.5	5.1	10.1	9.4	
3N7E29F1–SW11A	10.9	8.0	16.3	8.6	24.0	17.7	13.6	14.8	8.6	1.1	4.2	3.2	
3N7E32D2–SW12A	11.3	15.6	13.8	12.9	9.9	7.9	12.9	16.0	17.5	10.3	10.8	9.1	1,621.3
Total													3,548.1
							Mainside subbasin						
Golf course well	0.0	0.0	0.0	58.6	82.9	86.2	87.5	82.9	82.9	58.6	0.0	0.0	539.5
							Mesquite subbasin						
City of Twentynine Palms well	65.1	60.4	70.3	68.3	73.3	76.4	82.8	82.8	82.8	70.3	61.7	59.8	853.8

construction activities, and cooling) (table 9). In Scenario 3, about 10 percent of this nonpotable water demand was assumed to be met by pumpage from wells CW1 and CW2 in the Deadman subbasin and about 90 percent was from the six nonpotable supply wells in the Mainside subbasin (MS1–6). Pumpage from wells of the Surprise Spring subbasin was reduced by 610 acre-ft/yr compared to 2007 conditions in the inner hydrogeologic zone.

Table 13 shows the monthly pumpage distribution by well for Scenario 3. The total projected cumulative pumpage for Scenario 3 is about 49,420 acre-ft, the same as Scenario 2 (fig. 31). The pumpage from the inner hydrogeologic zone was reduced from about 1,300 acre-ft/yr in Scenario 2 to about 690 acre-ft/yr in Scenario 3 (tables 12, 13). The nonpotable pumpage from the Deadman subbasin (62 acre-ft/yr) was distributed evenly between CW1 and CW2 in the Deadman subbasin (table 13). The nonpotable pumpage from the Mainside subbasin not used on the golf course (about 550 acre-ft/yr) was distributed evenly between the six simulated nonpotable supply wells in the subbasin (MS1–6). The location of Scenario 3 pumping wells is shown on figure 32C (presented lated in the report).

Scenario 4

Scenario 4 assumes the same water-conservation plan as Scenarios 2 and 3 and the same reduction in potable water use as Scenario 3, and further reduces pumpage from Surprise Spring subbasin by blending the maximum amount of nonpotable water from Deadman subbasin with the minimum amount of potable water from the Surprise Spring subbasin needed to meet Federal and State drinking-water standards and water demand.

The proportions of pumpage from Surprise Spring subbasin (X) and the Deadman subbasin (Y) that when blended will meet Federal and State drinking water MCLs for a particular constituent i were estimated by simultaneously solving the following mass-balance equations for X and Y

$$XC_{x,i} + YC_{y,i} \leq (X + Y)C_{mcl,i} \qquad (1)$$

$$X + Y = 1, \qquad (2)$$

where $C_{x,i}$ is the average maximum concentration of constituent i (for example, concentration of arsenic, fluoride, or total dissolved solids) in groundwater from selected wells in the Surprise Spring subbasin, $C_{y,i}$ is the average maximum concentration of the same constituent in groundwater from selected wells in the Deadman subbasin, and $C_{mcl,i}$ is the MCL or secondary drinking-water standard for constituent i. For the Surprise Spring subbasin, $C_{x,i}$ was calculated as the average of the historical maximum concentrations of all production wells in the subbasin and was 5.1 µg/L for arsenic, 0.7 mg/L for fluoride, and 200 mg/L for total dissolved solids (table 14). For the Deadman subbasin, $C_{y,i}$ was calculated as the average of the historical maximum concentrations of wells close to the proposed nonpotable sites in the middle hydrogeologic zone in the Deadman subbasin and was 25 µg/L for arsenic,

5.4 mg/L for fluoride, and 775.4 mg/L for total dissolved solids (table 14). The USEPA MCL for arsenic is 10 µg/L, the USEPA MCL for fluoride is 4.0 mg/L, and the USEPA National Secondary Drinking Water Regulation for total dissolved solids is 500 mg/L (U.S. Environmental Protection Agency, 2010a,b).

Solution of equations (1) and (2) indicates that the maximum percentage of non-potable pumpage from the Deadman subbasin (Y) was 25 percent for arsenic, 70 percent for fluoride, and 50 percent for total dissolved solids concentrations (table 14). These results indicate that arsenic is the limiting constituent for blending. The total MCAGCC potable water demand for Scenario 4 was about 2,940 acre-ft/yr; therefore, about 2,205 acre-ft/yr (75 percent) was simulated as pumpage from the Surprise Spring subbasin and about 735 acre-ft/yr (25 percent) was simulated as pumpage from the Deadman subbasin. Projected annual pumpage for the Mainside and Mesquite subbasin is the same as Scenario 3.

Table 15 shows the monthly pumpage distribution by well for Scenario 4. The total projected cumulative pumpage for Scenario 4 is about 49,420 acre-ft, the same as Scenarios 2 and 3 (fig. 31). Location of Scenario 4 pumping wells is shown on figure 32D (presented lated in the report).

Simulation of Water-Management Scenarios

The four water-management scenarios were simulated by using the calibrated regional groundwater- flow model. The simulation period was January 2008 through December 2017. Results of the simulations are presented as hydraulic-head change maps for 2008–2017 for model layer 2 (fig. 32). A positive hydraulic-head change indicates that hydraulic-head increased from 2008 to 2017 (water-level rising) and a negative hydraulic-head change indicates that hydraulic head decreased from 2008–2017 (water-level declining). Hydrographs showing the simulated heads from 2008 to 2017 at model calibration wells are presented in appendix A for each of the scenarios.

Scenario 1

Scenario 1 simulated a 2-percent annual increase in pumpage from Surprise Spring subbasin from 2008 to 2017. The increased pumpage caused hydraulic head to decline by about 60 ft in the inner zone of the Surprise Spring (fig. 32A; appendix A, fig. A4). Pumping from the golf course well caused hydraulic head to decline about 50 ft in the Mainside subbasin near the well. Pumping from the City of Twentynine Palms well caused hydraulic-heads to decline by about 5 ft or less throughout a large part of the Mesquite subbasin.

Scenario 2

Scenario 2 assumes a water-conservation plan that will maintain essentially the same quantity of the MCAGCC pumpage as in the last year of the transient simulation (2007) except that about 540 acre-ft/yr of golf course pumpage was

Table 13. Monthly pumpage distribution by well for water-management scenario 3 for the Twentynine Palms area, California.

[All values are in acre-feet]

Well name	January	Febuary	March	April	May	June	July	August	September	October	November	December	Subtotal
						Surprise Spring subbasin							
						Inner zone							
2N7E03A1–SW3A	12.6	13.2	15.7	19.1	24.0	10.8	24.0	31.3	22.6	24.9	18.6	13.6	
3N7E35P2–SW4A	12.6	13.2	15.7	19.1	24.0	10.8	24.0	31.3	22.6	24.9	18.6	13.6	
2N7E02D1–SW5A	12.6	13.2	15.7	19.1	24.0	10.8	24.0	31.3	22.6	24.9	18.6	13.6	690.8
						Middle zone							
2N7E03B1–SW2A	8.1	2.7	22.6	26.9	27.3	27.8	31.2	16.9	19.3	12.5	9.3	6.1	
2N7E03E1–SW6A	23.0	24.6	33.0	13.2	40.0	39.1	47.0	45.4	47.4	44.0	35.0	22.7	624.9
						Outer zone							
3N7E28R1–SW7A	32.1	31.3	14.8	18.4	41.3	47.9	37.7	55.5	40.5	30.6	39.7	33.3	
3N7E32J1–SW8A	14.1	12.0	12.5	34.6	23.0	21.6	36.6	30.6	22.9	9.8	15.5	7.3	
3N7E29R1–SW9A	24.6	38.7	36.3	60.8	45.0	56.0	66.2	66.7	45.4	29.7	37.4	42.7	
3N7E28D1–SW10A	1.9	12.8	6.1	5.3	15.7	17.8	15.1	20.6	9.5	5.1	10.1	9.4	
3N7E29F1–SW11A	10.9	8.0	16.3	8.6	24.0	17.7	13.6	14.8	8.6	1.1	4.2	3.2	
3N7E32D2–SW12A	11.3	15.6	13.8	12.9	9.9	7.9	12.9	16.0	17.5	10.3	10.8	9.1	1,621.3
Total													2,937.1
						Deadman subbasin							
CW1	0.2	0.2	0.2	4.2	4.4	4.2	4.4	4.4	4.2	4.4	0.2	0.2	
CW2	.2	.2	.2	4.2	4.4	4.2	4.4	4.4	4.2	4.4	.2	.2	62.4
						Mainside subbasin							
Golf course well	0.0	0.0	0.0	58.6	82.9	86.2	87.5	82.9	82.9	58.6	0.0	0.0	
MS1	3.9	4.8	6.9	8.6	10.5	11.4	11.5	10.5	8.6	6.6	4.5	3.7	
MS2	3.9	4.8	6.9	8.6	10.5	11.4	11.5	10.5	8.6	6.6	4.5	3.7	
MS3	3.9	4.8	6.9	8.6	10.5	11.4	11.5	10.5	8.6	6.6	4.5	3.7	
MS4	3.9	4.8	6.9	8.6	10.5	11.4	11.5	10.5	8.6	6.6	4.5	3.7	
MS5	3.9	4.8	6.9	8.6	10.5	11.4	11.5	10.5	8.6	6.6	4.5	3.7	
MS6	3.9	4.8	6.9	8.6	10.5	11.4	11.5	10.5	8.6	6.6	4.5	3.7	1,088.2
						Mesquite subbasin							
City of Twentynine Palms well	65.1	60.4	70.3	68.3	73.3	76.4	82.8	82.8	82.8	70.3	61.7	59.8	853.8

Table 14. Historical maximum concentrations listed in the National Water Information System for arsenic (As), fluoride (F), and total dissolved solids in the Surprise Spring and Deadman subbasins, Federal drinking water standards, and the calculated percentage of groundwater pumped from each subbasin to meet drinking water standards in a blended water supply in the Twentynine Palms area, California.

[Data are from the U.S. Geological Survey National Water Inventory System database. NA, not available]

Well name	Well depth, in feet	Maximum As concentration, in micrograms per liter	Maximum F concentration, in milligrams per liter	Maximum total dissolved solids concentration, in milligrams per liter
Surprise Spring subbasin wells				
2N/7E–03E1	510	2.7	0.6	209
2N/7E–02D1	532	13.8	.9	175
2N/7E–03B1	700	2.5	.8	206
2N/7E–03A1	550	10.0	1.0	194
2N/7E–03A2	NA	9.3	1.1	170
3N/7E–35P2	593	10.0	.9	180
3N/7E–32J1	600	4.3	.6	177
3N/7E–32D2	570	3.0	.6	182
3N/7E–28R1	600	2.0	.5	234
3N/7E–29R1	600	2.3	.7	282
3N/7E–29F1	600	3.2	.4	189
3N/7E–28D1	618	2.0	1.2	207
3N/7E–28D2	600	.7	.4	195
Average concentration of Surprise Spring subbasin		5.1	.7	200.0
Deadman subbasin wells				
2N/8E–04L1	591	6.0	4.3	890
2N/8E–04L2	380	21.0	2.2	906
2N/8E–04L3	285	86.0	2.6	995
3N/8E–31J1	390	4.0	1.2	537
3N/8E–31J2	320	21.0	5.5	320
3N/8E–31J3	250	69.0	2.7	264
3N/8E–28P1	395	1.0	7.2	1,060
3N/8E–28P2	180	7.0	11.1	1,030
3N/8E–28P3	85	15.0	5.5	944
3N/7E–27H1	587	20.0	5.4	578
3N/7E–36K1	777	NA	10.0	898
3N/7E–36G1	399	NA	9.6	315
3N/8E–29L1	590	NA	6.0	1,180
3N/8E–34D1	396	NA	4.5	938
3N/8E–33B1	512	NA	4.0	NA
3N/8E–29C1	800	NA	5.0	NA
3N/7E–13N1	487	NA	2.6	NA
3N/8E–17L1	456	NA	4.0	NA
2N/8E–07K1	525	NA	10.0	NA
Average concentration in Deadman subbasin		25.0	5.4	775.4
U.S. Environmental Protection Agency drinking water standards		10	4.0	500
Calculated percentage of water from Surprise Spring subbasin (X) [1]		75	30	50
Calculated percentage of water from the Deadman subbasin		25	70	50

[1] Calculated on the basis of the mass-balance equations presented in the "Description of Model Scenarios" section of this report.

Table 15. Monthly pumpage distribution by well for water-management scenario 4 for the Twentynine Palms area, California.

[All values are in acre-feet]

Well name	January	February	March	April	May	June	July	August	September	October	November	December	Subtotal
Surprise Spring subbasin													
Inner zone													
2N7E03A1-SW3A	9.4	9.9	11.8	14.3	18.0	8.1	18.0	23.5	16.9	18.7	13.9	10.2	
3N7E35P2-SW4A	9.4	9.9	11.8	14.3	18.0	8.1	18.0	23.5	16.9	18.7	13.9	10.2	
2N7E02D1-SW5A	9.4	9.9	11.8	14.3	18.0	8.1	18.0	23.5	16.9	18.7	13.9	10.2	518.1
Middle zone													
2N7E03B1-SW2A	6.1	2.0	16.9	20.2	20.5	20.8	23.4	12.7	14.5	9.4	6.9	4.6	
2N7E03E1-SW6A	17.3	18.4	24.8	9.9	30.0	29.3	35.3	34.0	35.6	33.0	26.3	17.0	468.7
Outer zone													
3N7E28R1-SW7A	24.1	23.5	11.1	13.8	31.0	35.9	28.3	41.6	30.4	23.0	29.7	25.0	
3N7E32J1-SW8A	10.6	9.0	9.4	25.9	17.2	16.2	27.5	23.0	17.2	7.3	11.7	5.5	
3N7E29R1-SW9A	18.4	29.0	27.2	45.6	33.8	42.0	49.7	50.1	34.0	22.2	28.0	32.0	
3N7E28D1-SW10A	1.4	9.6	4.6	3.9	11.7	13.3	11.3	15.4	7.2	3.8	7.6	7.0	
3N7E29F1-SW11A	8.2	6.0	12.2	6.5	18.0	13.3	10.2	11.1	6.4	.8	3.2	2.4	
3N7E32D2-SW12A	8.5	11.7	10.4	9.7	7.5	5.9	9.7	12.0	13.1	7.7	8.1	6.8	1,216.0
Total													**2,202.8**
Deadman subbasin													
CW1	20.7	23.4	25.5	34.0	41.6	37.7	45.9	49.4	39.1	31.6	27.4	22.0	
CW2	20.7	23.4	25.5	34.0	41.6	37.7	45.9	49.4	39.1	31.6	27.4	22.0	796.7
Mainside subbasin													
Golf course well	0.0	0.0	0.0	58.6	82.9	86.2	87.5	82.9	82.9	58.6	0.0	0.0	
MS1	3.9	4.8	6.9	8.6	10.5	11.4	11.5	10.5	8.6	6.6	4.5	3.7	
MS2	3.9	4.8	6.9	8.6	10.5	11.4	11.5	10.5	8.6	6.6	4.5	3.7	
MS3	3.9	4.8	6.9	8.6	10.5	11.4	11.5	10.5	8.6	6.6	4.5	3.7	
MS4	3.9	4.8	6.9	8.6	10.5	11.4	11.5	10.5	8.6	6.6	4.5	3.7	
MS5	3.9	4.8	6.9	8.6	10.5	11.4	11.5	10.5	8.6	6.6	4.5	3.7	
MS6	3.9	4.8	6.9	8.6	10.5	11.4	11.5	10.5	8.6	6.6	4.5	3.7	1,088.2
Mesquite subbasin													
City of Twentynine Palms well	65.1	60.4	70.3	68.3	73.3	76.4	82.8	82.8	82.8	70.3	61.7	59.8	853.8

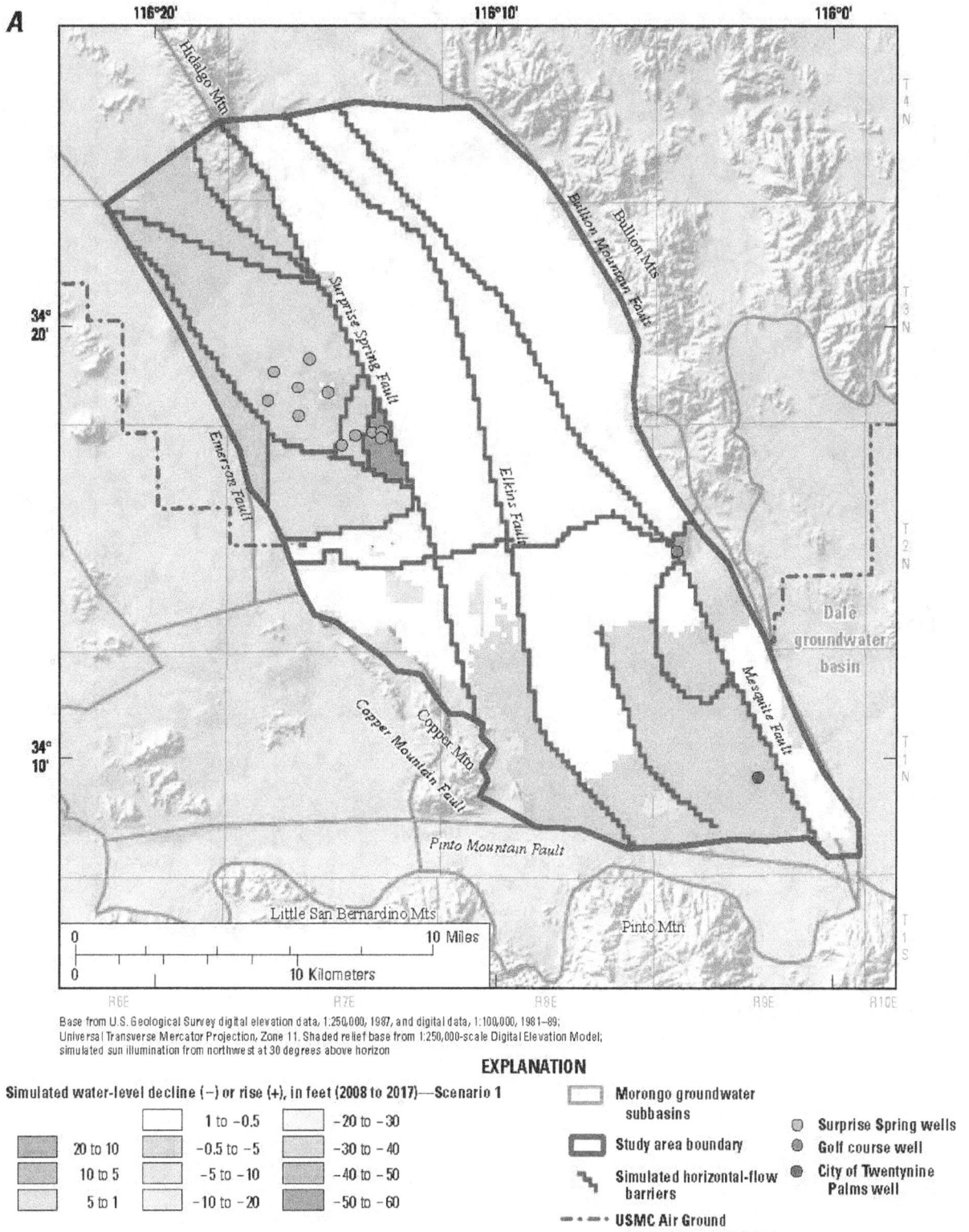

Figure 32.   Simulated change in hydraulic head in model-layer 2 for water-management scenarios (*A*) 1, (*B*) 2, (*C*) 3, and (*D*) 4, in the Twentynine Palms area, California, 2008–2017.

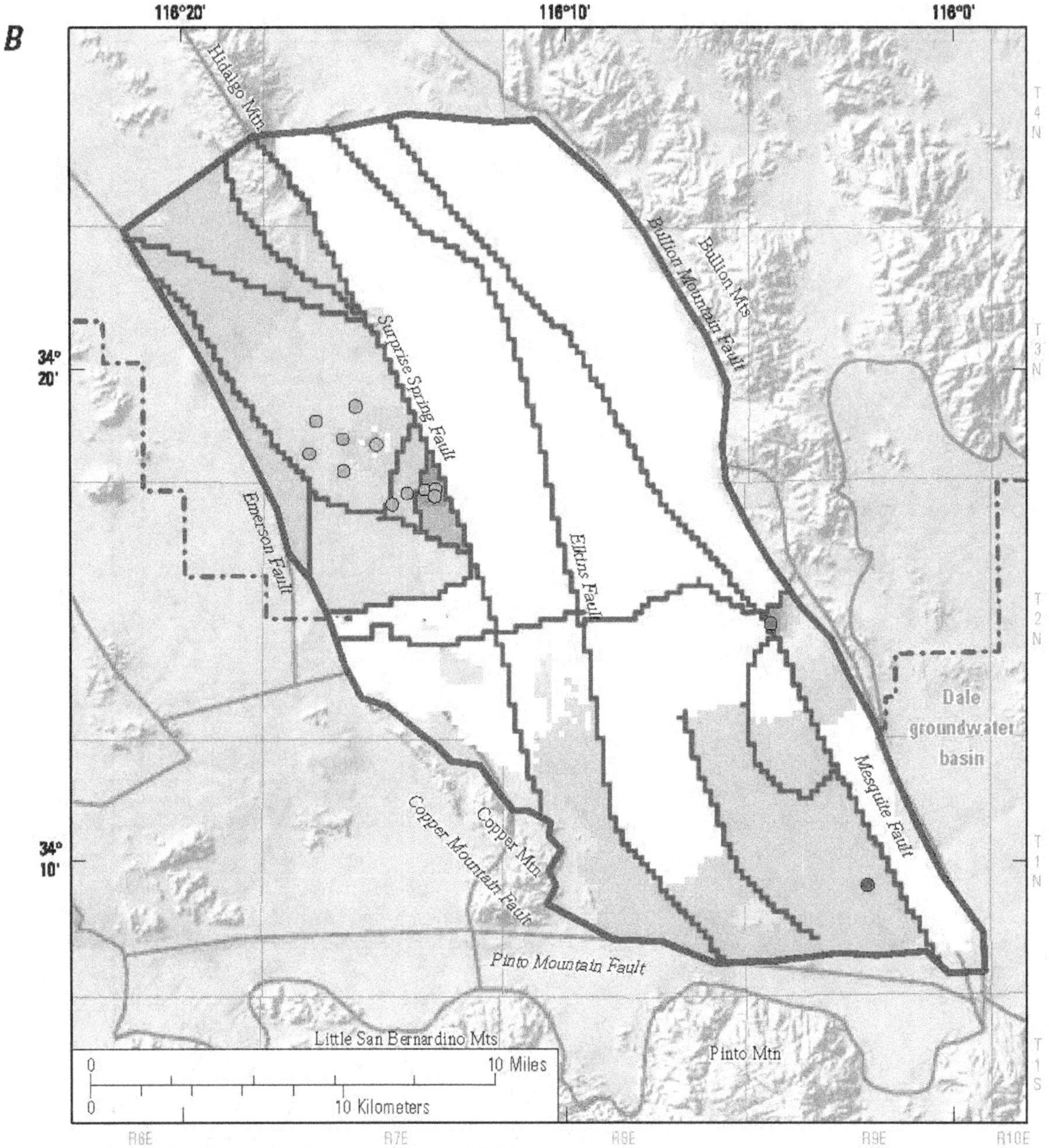

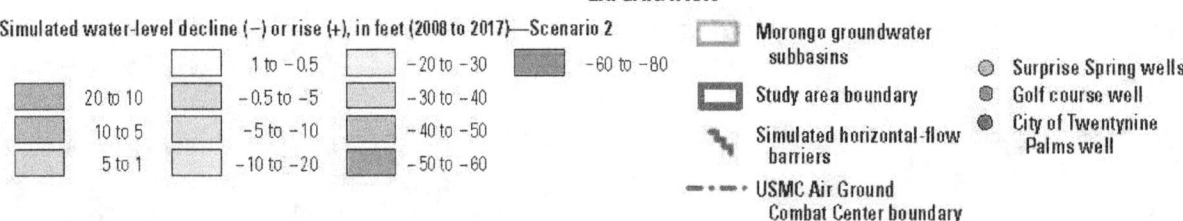

Base from U.S. Geological Survey digital elevation data, 1:250,000, 1987, and digital data, 1:100,000, 1981–89;
Universal Transverse Mercator Projection, Zone 11. Shaded relief base from 1:250,000-scale Digital Elevation Model;
simulated sun illumination from northwest at 30 degrees above horizon

EXPLANATION

Simulated water-level decline (−) or rise (+), in feet (2008 to 2017)—Scenario 2

20 to 10	1 to −0.5
10 to 5	−0.5 to −5
5 to 1	−5 to −10
	−10 to −20

−20 to −30	−60 to −80
−30 to −40	
−40 to −50	
−50 to −60	

Morongo groundwater subbasins

Study area boundary

Simulated horizontal-flow barriers

— · · — · USMC Air Ground Combat Center boundary

Surprise Spring wells

Golf course well

City of Twentynine Palms well

Figure 32. Continued.

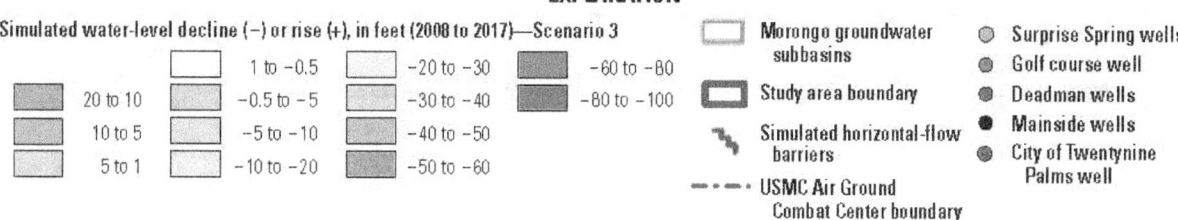

C

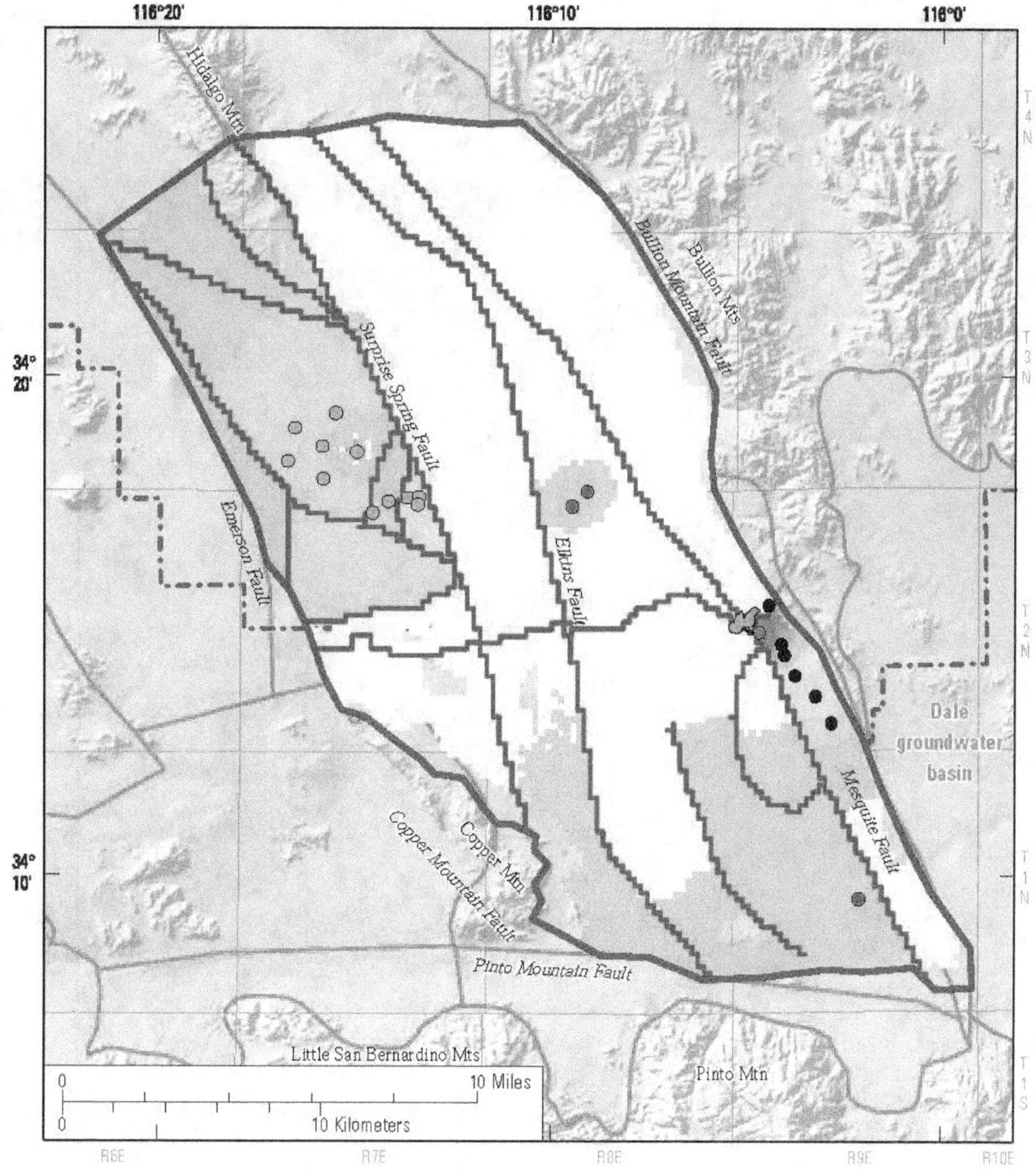

116°20'   116°10'   116°0'

Hidalgo Mtn

Bullion Mountain Fault

Bullion Mts

34°
20'

Surprise Spring Fault

Emerson Fault

Elkins Fault

T 4 N

T 3 N

T 2 N

Dale
groundwater
basin

Mesquite Fault

34°
10'

Copper Mtn

Copper Mountain Fault

T 1 N

Pinto Mountain Fault

Little San Bernardino Mts

Pinto Mtn

T 1 S

0   10 Miles

0   10 Kilometers

R6E   R7E   R8E   R9E   R10E

Base from U.S. Geological Survey digital elevation data, 1:250,000, 1987, and digital data, 1:100,000, 1981–89;
Universal Transverse Mercator Projection, Zone 11. Shaded relief base from 1:250,000-scale Digital Elevation Model;
simulated sun illumination from northwest at 30 degrees above horizon

EXPLANATION

Simulated water-level decline (−) or rise (+), in feet (2008 to 2017)—Scenario 3

	1 to −0.5	−20 to −30	−60 to −80
20 to 10	−0.5 to −5	−30 to −40	−80 to −100
10 to 5	−5 to −10	−40 to −50	
5 to 1	−10 to −20	−50 to −60	

Morongo groundwater
  subbasins

Study area boundary

Simulated horizontal-flow
  barriers

USMC Air Ground
  Combat Center boundary

○ Surprise Spring wells
◐ Golf course well
◑ Deadman wells
● Mainside wells
○ City of Twentynine
    Palms well

Figure 32.   Continued.

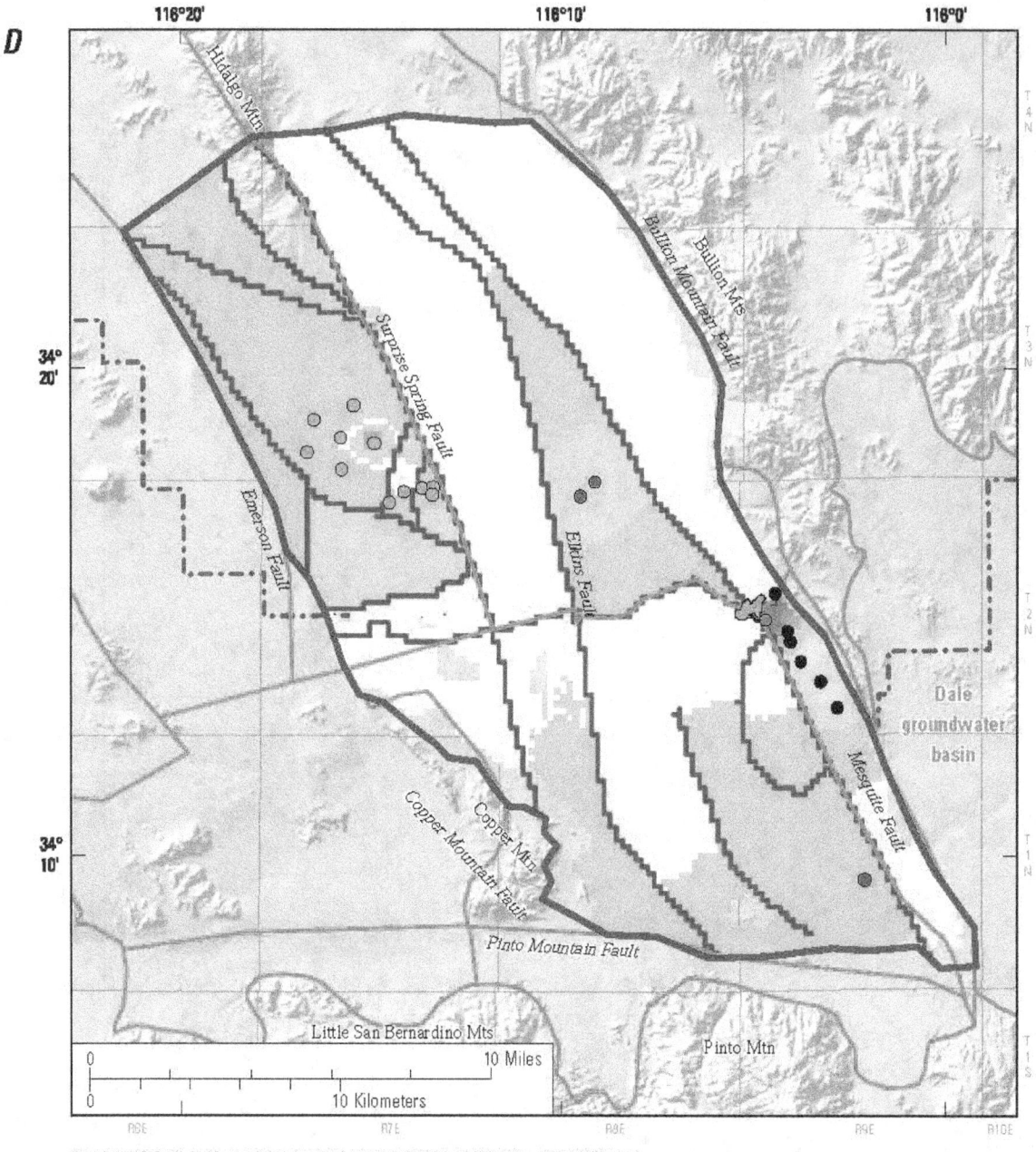

Base from U.S. Geological Survey digital elevation data, 1:250,000, 1987, and digital data, 1:100,000, 1981–89;
Universal Transverse Mercator Projection, Zone 11. Shaded relief base from 1:250,000-scale Digital Elevation Model;
simulated sun illumination from northwest at 30 degrees above horizon

EXPLANATION

Simulated water-level decline (−) or rise (+), in feet (2008 to 2017)—Scenario 4

26 to 20	1 to −0.5	−20 to −30	−60 to −80
20 to 10	−0.5 to −5	−30 to −40	−80 to −100
10 to 5	−5 to −10	−40 to −50	
5 to 1	−10 to −20	−50 to −60	

- Morongo groundwater subbasins
- Study area boundary
- Simulated horizontal-flow barriers
- USMC Air Ground Combat Center boundary

- Surprise Spring wells
- Golf course well
- Deadman wells
- Mainside wells
- City of Twentynine Palms well

Figure 32. Continued.

simulated in the Mainside subbasin. Simulation results suggest that continuing 2007 pumpage in the Surprise Spring subbasin may cause hydraulic heads to decline more than 50 ft in the inner hydrogeologic zone of the subbasin by 2017 (fig. 32*B*; appendix A, fig. A4). Hydraulic-head changes in the remainder of the model domain were similar to those in Scenario 1.

Scenario 3

Scenario 3 assumes the same water-conservation plan as in Scenario 2 and adds measures to replace 610 acre-ft/yr of potable water pumped from the inner hydrogeologic zone of the Surprise Spring subbasin with nonpotable water pumped from the Deadman and Mainside subbasins. Pumpage would be reduced by about 50 percent in the inner zone of the Surprise Spring subbasin and about 17 percent in the entire subbasin (tables 12 and 13). Reducing pumpage from the inner hydrogeologic zone caused the simulated hydraulic head to decline 10–20 ft in the inner zone by 2017, about 40 ft less than the hydraulic-head decline simulated in Scenario 1 (figs. 32*A*,*C*). Reducing pumpage from the inner zone of the Surprise Spring subbasin has a large impact on simulated hydraulic-heads within this zone because the storage volume of the zone is relatively small and the zone is bounded by low-permeability faults. Redistributing pumpage from the Surprise Spring subbasin to the Deadman and the Mainside subbasins resulted in a maximum simulated hydraulic-head decline of about 5 ft in the Deadman subbasin and more than 60 ft near the golf course in the Mainside subbasin (fig. 32*C*).

Scenario 4

Scenario 4 assumes the same water-conservation plan as Scenarios 2 and 3 and the same reduction in potable water use as Scenario 3 (about 610 acre-ft/yr), and further reduces pumpage from the Surprise Spring subbasin by about 25 percent (about 735 acre-ft/yr). The 25 percent reduction from the Surprise Spring subbasin is offset by pumping nonpotable water from the Deadman subbasin to blend with potable water from the Surprise Spring subbasin while still meeting water demand and Federal drinking water standards. Reducing pumpage from the Surprise Spring subbasin in a large amount decreased or reversed previously simulated hydraulic-head declines throughout the Surprise Spring subbasin during the 10-year simulation (fig. 32*D*; appendix A, figs. A1–A5). Simulated changes in hydraulic-head were about -0.5 to -10 ft in the inner zone, -5 to 5 ft in the middle zone, and -10 to 10 ft in the outer zone.

Redistributing pumping from the Surprise Spring subbasin to the Deadman and the Mainside subbasins resulted in a maximum simulated hydraulic-head decline of about 10 ft in the Deadman subbasin and more than 60 ft near the golf course in the Mainside subbasin (fig. 32*D*). Increasing pumpage in the Deadman subbasin increased the area where simulated hydraulic head declined in the middle hydrogeologic zone of the Deadman subbasin. The Elkins Fault to the west, the Mesquite Fault to the east, and the Transverse Arch

to the south control the areal extent of the simulated hydraulic-head decline.

Summary of Water-Management Scenarios

Simulation results of the four water-management scenarios show the important effects of reducing groundwater pumpage from the Surprise Spring subbasin. If groundwater pumpage increases by 2 percent annually in the subbasin (Scenario 1), hydraulic heads are estimated to decline about 60 ft in the inner hydrogeologic zone of the subbasin from 2008 to 2017. If pumpage from the Surprise Spring subbasin is maintained at 2007 rates (Scenario 2), hydraulic heads are estimated to decline about 50 ft. These simulated hydraulic-head declines are in addition to the more than 190-ft decline in water levels that occurred from 1953 to 2007.

The results of Scenarios 3 and 4 show that reducing groundwater pumpage in the Surprise Spring subbasin, especially in the inner hydrogeologic zone of the subbasin, would prolong the limited supply of potable groundwater in the upper aquifer of the subbasin. Reducing the 2007 groundwater pumpage from the Surprise Spring subbasin by about 38 percent (about 1,345 acre-ft/yr), with about 60 percent (about 784 acre-ft/yr) of the reduction in the inner zone, substantially decreased or reversed simulated hydraulic-head declines in the subbasin (Scenario 4). Reducing pumpage from the inner hydrogeologic zone of the Surprise Spring subbasin has a significant effect on simulated hydraulic-heads within this zone because the inner zone has a relatively small storage volume and is bounded by low-permeability faults.

Redistributing about 15 percent (about 548 acre-ft/yr) of the 2007 groundwater pumpage from the Surprise Spring to the Mainside subbasins (Scenarios 3 and 4) resulted in more than 60 ft of simulated declines in hydraulic head in the Mainside subbasin because of the narrow width of the subbasin (fig. 32*C*), existing pumping from the golf course well, and low estimated hydraulic conductivity of the aquifer (4 ft/d for layer 2, table 3). However, redistributing about 22 percent (about 800 acre-ft/yr) of the 2007 groundwater pumpage from the Surprise Spring subbasin to the Deadman subbasin (Scenario 4) resulted in only 5–10 ft of hydraulic-head decline near the simulated pumping wells in the middle hydrogeologic zone of the Deadman subbasin (fig. 32*D*). The areal extent of this head decline is controlled by the Elkins Fault to the west, the Mesquite Fault to the east, and the Transverse Arch to the south. Because of the proximity of the ecologically sensitive Mesquite Lake (dry) (fig. 2) to the proposed pumping locations, it will be important to monitor water levels on both sides of the Transverse Arch to determine if the geologic structure is an effective barrier to groundwater flow.

All of the water-management scenarios indicated that pumpage from the City of Twentynine Palms well will cause the simulated hydraulic head to decline slightly (less than 5 ft) throughout a large part of the Mesquite subbasin. It is important to note that the simulated hydraulic-head decline, although small, affects the Mesquite Lake (dry) area (figs. 2, 31). Declines in hydraulic head in the Mesquite subbasin will eventually decrease the amount of natural groundwater discharge from the subbasin. Because phreatophytes (mostly mesquite) in the Mesquite Dry Lake area use groundwater as a source of water, declining groundwater levels may adversely affect the health and survivability of the phreatophytes.

The costs associated with developing and operating water-supply systems that would be needed to implement the water-management scenarios were not evaluated for this study. Conservation measures assumed in all scenarios except Scenario 1 may require purchasing new water-efficient appliances. Development of nonpotable water in the Deadman and the Mainside subbasins for Scenarios 3 and 4 would involve drilling or rehabilitating wells, and Scenario 4 would require building new pipelines and storage tanks to blend water from the Surprise Spring and the Deadman subbasins. To identify the best water-management strategy, both hydrologic and economic factors need to be considered. The water-management scenarios simulated for this study demonstrate how the calibrated regional model can be used to evaluate the hydrologic factors of various water-management strategies. A simulation-optimization model could be developed to quantitatively evaluate the hydrologic and economic factors of alternative water-management strategies.

Summary and Conclusions

The U.S. Marine Corps Air Ground Combat Center (MCAGCC), the largest air/ground combat training center of the U.S. Marine Corps, is located in a remote desert area. Historically, the MCAGCC has relied on groundwater pumped from the Surprise Spring subbasin to provide all of its water supply. Groundwater from other nearby subbasins (Deadman, Mainside, and Mesquite) contains unacceptably high concentrations of fluoride, arsenic, and (or) dissolved solids, making it unsuitable for potable uses without treatment. The potable groundwater supply in Surprise Spring subbasin has diminished because of pumpage-induced overdraft and more restrictive Federal drinking water standards for arsenic concentrations. The MCAGCC needs to develop a plan for thoroughly understanding its available groundwater resources and to establish a long-term strategy for regional water-resource development to ensure the future viability of water supply at the base.

The purpose of this study was to improve the understanding of the geohydrology of the Surprise Spring, Deadman, Mesquite, and Mainside subbasins of the Morongo groundwater basin and develop a regional groundwater-flow model of these subbasins to help manage the water resources of the region. The Morongo groundwater basin underlies a broad, eastward-sloping alluvial desert plain, almost completely surrounded by mountains and uplands. The mountain ranges and uplands consist of a nearly impermeable complex of igneous and metamorphic rocks of pre-Tertiary age. A gravity survey completed for the study area identified two extremely deep basins beneath the Deadman and the Mesquite Lakes (dry) where the depth to the basement complex is more than 16,000 ft. The depth to the basement complex at the Surprise Spring subbasin is relatively shallow, with a maximum depth of less than 4,300 ft.

The subbasins, from the bottom to the top, are filled with older sedimentary deposits of Tertiary age, older alluvial-fan deposits of Quaternary/Tertiary age, and alluvial and playa deposits of Quaternary age. The deposits are unconsolidated at land surface and become more consolidated with depth. The Tertiary sedimentary deposits fill most of the depth of the subbasins, with the Quaternary/Tertiary alluvial-fan deposits and Quaternary alluvial and playa deposits forming a thin crust overlying the Tertiary-age deposits. The older sedimentary deposits yield a small amount of water to wells, and that water commonly contains high concentrations of fluoride, arsenic, and dissolved solids. The older alluvial fan deposits are the principal water-bearing unit in the study area, and have a combined thickness of 250 to more than 1,000 ft. Despite the tremendous thickness of the sedimentary deposits in the study area, most of the area has limited groundwater resources because of the relatively thin layer of saturated alluvial fan deposits that yield water freely to wells.

The study area is dominated by extensive faulting and moderate to intense folding that has displaced or deformed the pre-Tertiary basement complex as well as Tertiary and Quaternary deposits. Many of these faults are barriers to the lateral movement of groundwater flow and form the boundaries of the groundwater subbasins.

Interpretations from lithologic and downhole geophysical logs were used to identify two aquifers (referred to as the upper and the middle aquifers) in the Quaternary/Tertiary alluvial fan deposits and a single aquifer (referred to as the lower aquifer) in the Tertiary older sedimentary deposits. The pre-Tertiary basement complex forms the base of the aquifer system. In general, wells perforated in the upper aquifer in the Surprise Spring subbasin yield more water than wells perforated in the middle and lower aquifers. The inner and middle hydrogeologic zones of the Surprise Spring subbasin have the greatest saturated thickness of the upper aquifer in the study area. The upper aquifer is thin or is unsaturated in the Deadman, Mesquite, and Mainside subbasins.

The principal recharge to the study area is groundwater underflow, which originated as runoff in the surrounding mountains, across the western and southern boundaries. Recharge from the infiltration of surface water within the study area is relatively small compared with recharge from groundwater underflow, and it occurs solely along washes in the Mesquite subbasin. Recharge from direct precipitation over the area is insignificant. Groundwater discharges naturally from the study area as spring flow, as groundwater underflow to downstream basins, and as water vapor to the atmosphere by transpiration of phreatophytes and direct evaporation from moist soil. Before groundwater development in the study area, groundwater recharge was equal to discharge, and both are estimated to have been about 1,010 acre-ft/yr.

The MCAGCC has been the primary user of groundwater in the study area since 1953. From 1953 to 2007, approximately 139,400 acre-ft of groundwater was pumped out of the Surprise Spring subbasin from 11 supply wells. During that period, groundwater pumpage from the Surprise Spring subbasin caused water levels to decline as much as 190 ft. Water-level measurements indicate that water levels started to decline in the subbasin near Surprise Spring almost immediately after the pumping initiated in 1953. Water levels in the Deadman, Mesquite, and Mainside subbasins have been relatively stable during this same period because groundwater pumpage in these subbasins has been minimal.

A regional-scale numerical groundwater-flow model was developed using MODFLOW–2000 for the Surprise Spring, Deadman, Mesquite, and Mainside subbasins to better understand the aquifer system used by the MCAGCC for its water supply and to use as a tool to help manage groundwater resources in the Twentynine Palms area. The model of the aquifer system has three layers. Layer 1 represents the upper aquifer, layer 2 represents the middle aquifer, and layer 3 represents the lower aquifer. The model was calibrated by using a trial-and-error process in which the initial estimates of the aquifer properties and groundwater underflow were iteratively adjusted to improve the match between simulated hydraulic heads and measured groundwater levels. Measured groundwater levels for the predevelopment period (before 1953) and for the period 1953 through 2007 were used to calibrate the groundwater-flow model for steady-state and transient groundwater conditions, respectively.

For the pre-development condition, model simulated steady-state recharge for the study area was about 980 acre-ft/yr; about 90 percent of the recharge occurred in the Mesquite subbasin. Simulated groundwater recharge in the heavily pumped Surprise Spring subbasin was only 110 acre-ft/yr. Most of the simulated steady-state discharge occurred as evapotranspiration at the Mesquite Lake (dry).

For the development conditions, about 145,450 acre-ft of groundwater was simulated as being pumped from the model domain during the transient simulation period (1953–2007): about 139,400 acre-ft from the Surprise Spring subbasin, about 2,380 acre-ft from the Deadman subbasin, and about 3,670 acre-ft from the Mesquite subbasin. The transient simulation suggests that almost all of the groundwater pumped in the Surprise Spring subbasin comes from groundwater storage. Groundwater pumpage and depletion of groundwater storage reached a maximum rate of about 4,200 acre-ft/yr in 2000. The simulated and measured water-level declines in the Surprise Spring are the result of this depletion in groundwater storage. The groundwater pumping and resulting groundwater-level declines have decreased the saturated thickness of the productive upper aquifer in the Surprise Spring subbasin almost 50 percent since 1953 in the inner hydrogeologic zone near the ancestral Surprise Spring.

The calibrated groundwater model was used to help evaluate the effects of water-management strategies being considered by the MCAGCC to meet the projected water demand at the base for 2008–2017. One of the primary goals is to replace groundwater pumpage of potable water in the Surprise Spring subbasin with groundwater pumpage of nonpotable water in the Deadman and Mainside subbasins.

Water-management strategies being considered by the MCAGCC were evaluated by using the calibrated regional groundwater-flow model to simulate four pumpage scenarios. Scenario 1 assumes a 20-percent increase in population, which approximately translates into a 2-percent annual increase in water demand and pumpage from 2008 to 2017. Scenario 2 assumes the same 20-percent increase in population and that the MCAGCC will institute a water-conservation plan that will conserve 2 percent of its potable water supply from Surprise Spring subbasin annually. The water conservation offsets the projected increase in the MCGACC potable water supply; thus maintaining the MCAGCC pumpage from the Surprise Spring subbasin at the 2007 level. Scenario 3 simulates the same water-conservation plan as simulated in Scenario 2 and also replaces 610 acre-ft/yr of water pumped from the inner hydrogeologic zone of the Surprise Spring subbasin with nonpotable water pumped from the Deadman and the Mainside subbasins. Scenario 4 assumes the same water-conservation plan as Scenario 2 and the same reduction in potable water use as Scenario 3 (about 610 acre-ft/yr) and also reduces pumpage from Surprise Spring subbasin by about 25 percent (about 735 acre-ft/yr). The 25 percent reduction from the Surprise Spring subbasin is offset by pumping nonpotable water from the Deadman subbasin to blend with potable water from the Surprise Spring subbasin while still meeting water demand and Federal drinking water standards.

Results of the four water-management scenarios show the importance of reducing groundwater pumpage from the Surprise Spring subbasin. For groundwater pumpage increasing by 2 percent annually in the Surprise Spring subbasin (Scenario 1), hydraulic heads were simulated to decline about 60 ft in the inner hydrogeologic zone of the subbasin by 2017. For pumpage from the Surprise Spring subbasin being maintained at 2007 rates (Scenario 2), hydraulic heads were simulated to decline about 50 ft. These simulated hydraulic-head declines are in addition to the more than 190-ft decline that occurred from 1953 to 2007.

The results of Scenarios 3 and 4 show that reducing groundwater pumpage in the Surprise Spring subbasin, especially in the inner hydrogeologic zone, would prolong the limited supply of potable groundwater in the upper aquifer of the subbasin. Reducing the total pumpage from the subbasin by 38 percent (about 1,345 acre-ft/yr), with about 60 percent (about 784 acre-ft/yr) of the reduction in the inner zone, substantially decreased or reversed simulated hydraulic-head declines in the subbasin (Scenario 4). Reducing pumpage from the inner hydrogeologic zone in the Surprise Spring subbasin has a significant effect on simulated hydraulic-heads within this zone because the zone has a relatively small storage volume and is compartmentalized by bounding low-permeability faults.

Redistributing about 15 percent of the 2007 groundwater pumpage (about 550 acre-ft/yr) from the Surprise Spring to the Mainside subbasin caused the simulated hydraulic head to decline more than 60 ft in the Mainside subbasin. However, redistributing about 22 percent of the 2007 pumpage (about 800 acre-ft/yr) from the Surprise Spring subbasin to the Deadman subbasin caused the simulated hydraulic head to decline only 5–10 ft in the Deadman subbasin. The areal extent of the simulated hydraulic-head decline in the Deadman subbasin is controlled by the Elkins Fault to the west, the Mesquite Fault to the east, and the Transverse Arch to the south. Because of the proximity of the ecologically sensitive Mesquite Lake (dry) to the proposed pumping locations, it will be important to monitor water levels on both sides of the Transverse Arch to determine if the geologic structure is an effective barrier to groundwater flow.

All of the water-management scenarios indicated that pumpage from the Twentynine Palms well will cause hydraulic heads to decline slightly (less than 5 ft) throughout a large part of the Mesquite subbasin. It is important to note that the simulated hydraulic-head decline, although small, impacts the Mesquite Lake (dry) area. Declines in hydraulic head in the Mesquite subbasin will eventually decrease the amount of natural groundwater discharge from the subbasin. Because phreatophytes (mostly mesquite) in the Mesquite Dry Lake area use groundwater as a major source of water, declining groundwater levels may adversely affect the health and survivability of the phreatophytes.

The costs associated with developing and operating the water-supply systems needed to implement the water-management scenarios were not evaluated for this study. Conservation measures required in all scenarios except Scenario 1 may require new water-efficient appliances. Developing nonpotable water in the Deadman and the Mainside basins simulated for Scenarios 3 and 4 will involve drilling or rehabilitating wells, and Scenario 4 will require building new pipelines and storage tanks to blend the Surprise Spring and the Deadman pumped water. To identify the best water-management strategy, hydrologic and economic factors need to be considered. The water-management scenarios simulated for this study demonstrate how the calibrated regional model can be utilized to evaluate the hydrologic factors of different water-management strategies. A simulation-optimization model could be developed to systematically evaluate the hydrologic and the economic aspects of different water-management strategies.

Acknowledgements

This report was completed as part of a cooperative study between the USGS and the MCAGCC. The authors are grateful for the funding provided by the MCAGCC to undertake this study and the cooperation given by personnel of the MCAGCC in providing hydrologic data and assistance in the successful completion of the study. The authors also thank the following USGS personnel for their contribution to the completion of this report: Dennis A. Clark for data collection, Brett Cox for geologic interpretation, Joseph Hevesi for estimation of natural recharge, Steven K. Predmore for GIS assistance, Larry G. Schneider for scientific illustration, and Mary A. Gibson for editorial assistance.

References

Akers, J.P., 1986, Geohydrology and potential for artificial recharge in the western part of the U.S Marine Corps Base, Twentynine Palms, California, 1982–83: U.S. Geological Survey Water-Resources Investigations Report 84–4119, 18 p.

California Irrigation Management Information System, 2002, California ETo zones map: California Department of Water Resources data available at *http://www.cimis.water.ca.gov/*.

Haley & Aldrich, Inc., 2001, Baseline report—Compilation of groundwater and well information within the Twentynine Palms Water District Area, v. I, II.

Halford, K.J., and Hanson, R.T., 2002, User guide for the drawdown-limited, multi-node well (MNW) package for the U.S. Geological Survey's modular ground-water flow model, versions MODFLOW–96 and MODFLOW–2000: U.S. Geological Survey Open-File Report 02–293, 33 p.

Harbaugh, A.W., 1990, A computer program for calculating subregional water budgets using results from the U.S. Geological Survey Modular Three-Dimensional Finite- Difference Ground-Water Flow Model: U.S. Geological Survey Open-File Report 90–392, 46 p.

Harbaugh, A.W., Banta, E.R., Hill, M.C., and McDonald, M.G., 2000, MODFLOW-2000, the U.S. Geological Survey modular ground-water model—User guide to modularization concepts and the Ground-Water Flow Process: U.S. Geological Survey Open-File Report 00–92, 121 p.

Hevesi, J.A., Flint, A.L., and Flint, L.E., 2003, Simulation of net infiltration and potential recharge using the distributed-parameter watershed model, INFILv3, of the Death Valley Region, Nevada and California: U.S. Geological Survey Water-Resources Investigations Report 03–4090, 161 p.

Hill, M.C., Banta, E.R., Harbaugh, A.W., and Anderman, E.R., 2000, MODFLOW-2000, the U.S. Geological Survey modular ground-water flow model—User guide to the observation, sensitivity, and parameter-estimation processes and three post-processing programs: U.S. Geological Survey Open-File Report 00–184, 209 p.

Hsieh, P.A., and Freckleton, J.R., 1993, Documentation of a computer program to simulate horizontal-flow barriers using the U.S. Geological Survey's modular three-dimensional finite-difference ground-water flow model: U.S. Geological Survey Open-File Report 92–477, 32 p.

Izbicki, J.A., 2004, Source and movement of ground water in the western part of the Mojave Desert, Southern California, USA: U.S. Geological Survey Water-Resources Investigations Report 03–4313, 28 p.

Izbicki, J.A., and Michel, R.L., 2004, Movement and age of ground water in the western part of the Mojave Desert, Southern California, USA: U.S. Geological Survey Water-Resources Investigations Report 03–4314, 35 p.

Jachens, R.C., and Moring, B.C., 1990, Maps of the thickness of Cenozoic deposits and the isostatic residual gravity over basement for Nevada: U.S. Geological Survey Open-File Report 90–404, 15 p.

Kulongoski, J.T., Hilton, D.R., and Izbicki, J.A., 2005, Source and movement of helium in the eastern Morongo groundwater basin: The influence of regional tectonics on crustal and mantle helium fluxes: Geochimica et Cosmochimica Acta, v. 69, no. 15, p. 3857–3872.

Lewis, R.E., 1972, Ground-water resources of the Yucca Valley-Joshua Tree area, San Bernardino County, California: U.S. Geological Survey Open-File Report, 72–234, 51 p.

Lohman, S.W., 1972, Ground-water hydraulics: U.S. Geological Survey Professional Paper 708, 70 p.

Londquist, C.J., and Martin, Peter., 1991, Geohydrology and ground-water-flow simulation of the Surprise Spring Basin Aquifer System, San Bernardino County, California: U.S. Geological Survey Water-Resources Investigations Report 89–4099, 41 p.

Matti, J.C., and Morton, D.M., 1994, Preliminary geologic map of part of the U.S. Marine Corps Twentynine Palms Air-Ground Combat Center, California: U.S. Geological Survey Administrative Report USGS–WRG–94–010, 17 p., scale 1:24,000.

Matti, J.C., and Morton, D.M., 1995, Preliminary geologic map of the Deadman Lake SW quadrangle, U.S. Marine Corps Twentynine Palms Air-Ground Combat Center, California: U.S. Geological Survey Administrative Report USGS–WRG–95–050, scale 1:24,000.

McDonald, M.G., and Harbaugh, A.W., 1988, A modular three-dimensional finite-difference ground-water flow model: U.S. Geological Survey Techniques of Water-Resources Investigations, book 6, chap. A1, 586 p.

Mendez, G.O., and Christensen, A.H., 1997, Regional water table (1996) and ground-water-level changes in the Mojave River, the Morongo, and Fort Irwin ground-water basins, San Bernardino County, California: U.S. Geological Survey Water-Resource Investigations Report 97–4160, 34 p., map in pocket.

Moyle, W.R., Jr., 1984, Bouger gravity anomaly map of the Twentynine Palms Marine Corps Base and vicinity, California: U.S. Geological Survey Water Resource Investigations Report 84–4005, scale 1:62,500.

Nishikawa, Tracy, Izbicki, J.A., Hevesi, J.A., Stamos, C.L., and Martin, Peter, 2004, Evaluation of recharge estimates and a ground-water flow model of the Joshua Tree area, San Bernardino County, California: U.S. Geological Survey Scientific Investigations Report 2004–5267, 115 p. Available at *http://pubs.usgs.gov/sir/2004/5267/*.

Riley, F.S., and Worts, G.F., Jr., 1952, Geologic reconnaissance and test-well drilling program, Marine Corps Training Center, Twentynine Palms, California: U.S. Geological Survey Open-File Report 98–166, 64 p. [retyped 2001].

Riley, F.S., and Worts, G.F., Jr., 1953, Geology and ground-water appraisal of the Twentynine Palms Marine Corps Training Center, California: U.S. Geological Survey Open-File Report 98–167, 113 p. [retyped 2001].

Roberts, C.W., Jachens, R.C., Katzenstein, A.M., Smith, G.A., and Johnson, R.U., 2002, Gravity map and data of the eastern half of the Big Bear Lake, 100,000 scale quadrangle, California, and analysis of the depths of several basins: U.S. Geological Survey Open-File Report 02–353. Available at *http://geopubs.wr.usgs.gov/open-file/of02-353/index.html*.

Roeloffs, E.A., Martin, P.M., Clark, D.A., and Pimentel, M.I., 2001, Fluid pressure changes in the Surprise Spring basin near Twenty-Nine Palms, California, induced by the 1992 Landers and 1999 Hector Mine earthquakes: Presented in American Geophysical Union Fall Meeting, San Francisco, December 10–14, 2001.

Schaefer, D.H., 1978, Ground-water resources of the Marine Corps Base, Twentynine Palms, San Bernardino County, California: U.S. Geological Survey Water-Resources Investigations Report 77–37, 29 p.

Smith, G.A., 2002, Regional water table (2000) and ground-water-level changes in the Mojave River and Morongo ground-water basins, San Bernardino County, California: U.S. Geological Survey Water-Resource Investigations Report 02-4277. Available at *http://pubs.usgs.gov/wri/wri024277/*.

Smith, G.A., and Pimentel, M.I., 2000, Regional water table (1998) and ground-water-level changes in the Mojave River and the Morongo ground-water basins, San Bernardino County, California: U.S. Geological Survey Water-Resources Investigations Report 00–4090, 107 p., map in pocket.

Smith, G.A., Stamos, C.L., and Predmore, S.K., 2004, Regional Water Table (2002) and Water-Level Changes in the Mojave River and Morongo Ground-Water Basins, Southwestern Mojave Desert, California: U.S. Geological Survey Scientific Investigations Report 2004–5081, 10 p.

Stamos, C.L., Huff, J.A., Predmore, S.K., and Clark, D.A., 2004, Regional water table (2004) and water-level changes in the Mojave River and Morongo ground-water basins, southwestern Mojave Desert, California: U.S. Geological Survey Scientific Investigations Report 2004–5187. Available at *http://pubs.usgs.gov/sir/2004/5187/*.

Stamos, C.L., McPherson, K.R., Sneed, Michelle, and Brandt, J.T., 2007, Water-level and land-subsidence studies in the Mojave River and Morongo ground-water basins: U.S. Geological Survey Scientific Investigations Report 2007–5097. Available at *http://pubs.usgs.gov/sir/2007/5097/*.

Thomasson, H.G., Jr., Olmsted, F.H., and LeRoux, E.F., 1960, Geology, water resources, and usable ground-water storage capacity of part of Solano County, California: U.S. Geological Survey Water-Supply Paper 1464, 693 p.

Trayler, C.R., and Koczot, K.M., 1995, Regional water table (1994) and ground-water-level changes in the Morongo basins, San Bernardino County, California: U.S. Geological Survey Water-Resource Investigations Report 95–4209, scale 1:125,000, 1 sheet.

Trexler, D.T., Flynn, T., and Ghusn, G., Jr., 1984, Drilling and thermal gradient measurements at U.S. Marine Corps Air-ground Combat Center, Twentynine Palms, California: Report by Division of Earth Sciences, Environmental Research Center, University of Nevada, Las Vegas. 79 p.

U.S. Environmental Protection Agency, 2010a, Drinking water contaminants: List of contaminants and their MCLs: U.S. Environmental Protection Agency. Available at *http://water.epa.gov/drink/contaminants/index.cfm#List*.

U.S. Environmental Protection Agency, 2010b, Drinking water contaminants: List of national secondary drinking water regulations: U.S. Environmental Protection Agency data available at *http://water.epa.gov/drink/contaminants/index.cfm#SecondaryList*.

Wagner, W.O., 1952, Hydrologic report on the ground-water resources of the area northwest of Twentynine Palms for the Marine Corps Training Center: mimeographed report, 22 p.

Wahler & Associates, 1983, City of Twentynine Palms, City demographic data, accessed on date at *http://www.ci.twentynine-palms.ca.us/City_Demographic_Data.63.0.html*.

Appendix A.   Long-Term Water-Level Hydrographs and Model-Simulated Hydraulic Heads at Selected Wells in the Twentynine Palms Area, California, 1953–2017

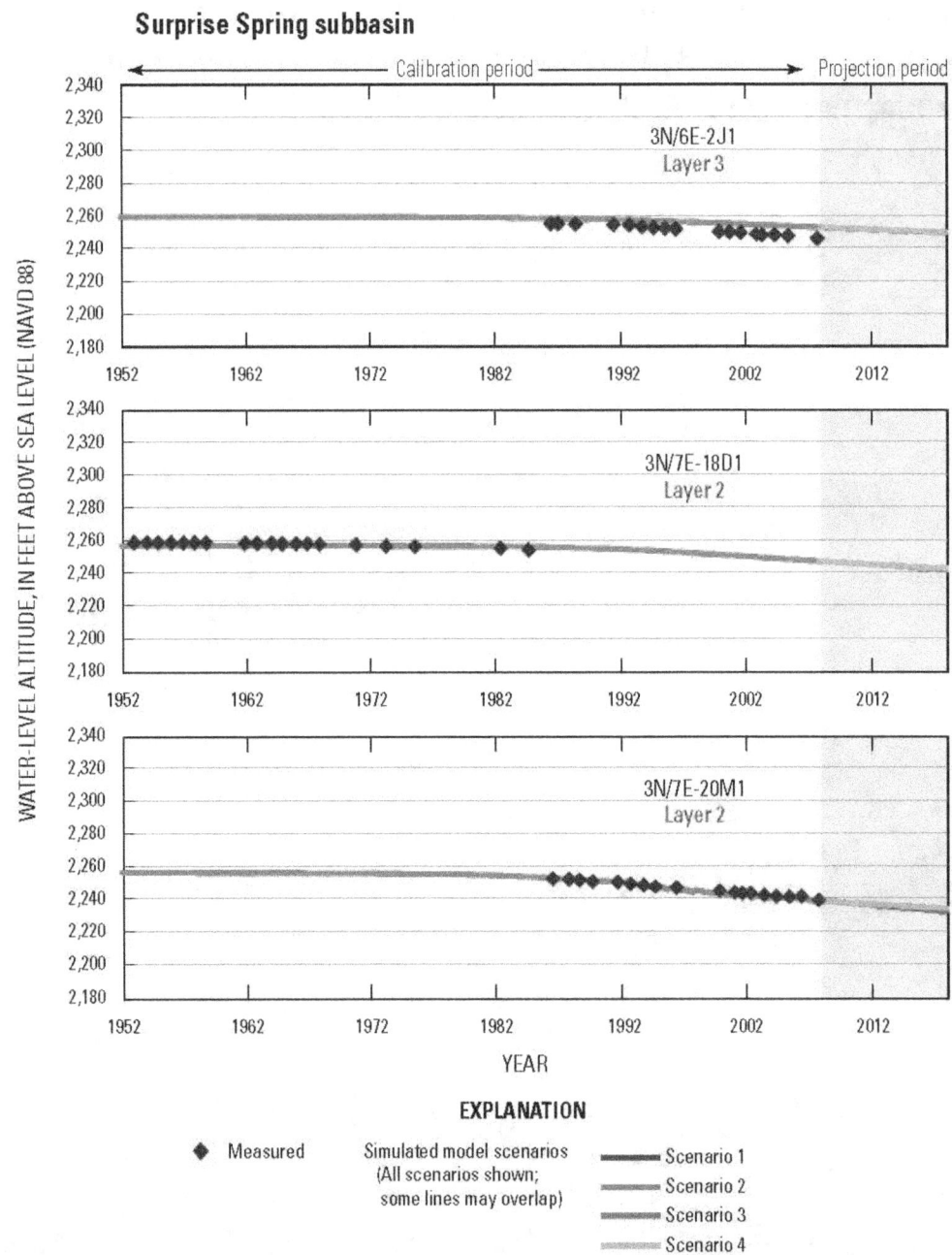

Figure A1   Long-term water-level hydrographs and model simulated hydraulic heads at wells 3N/6E-2J1, 3N/7E-18D1, and 3N/7E-20M1, Twentynine Palms area, California, 1953-2017.

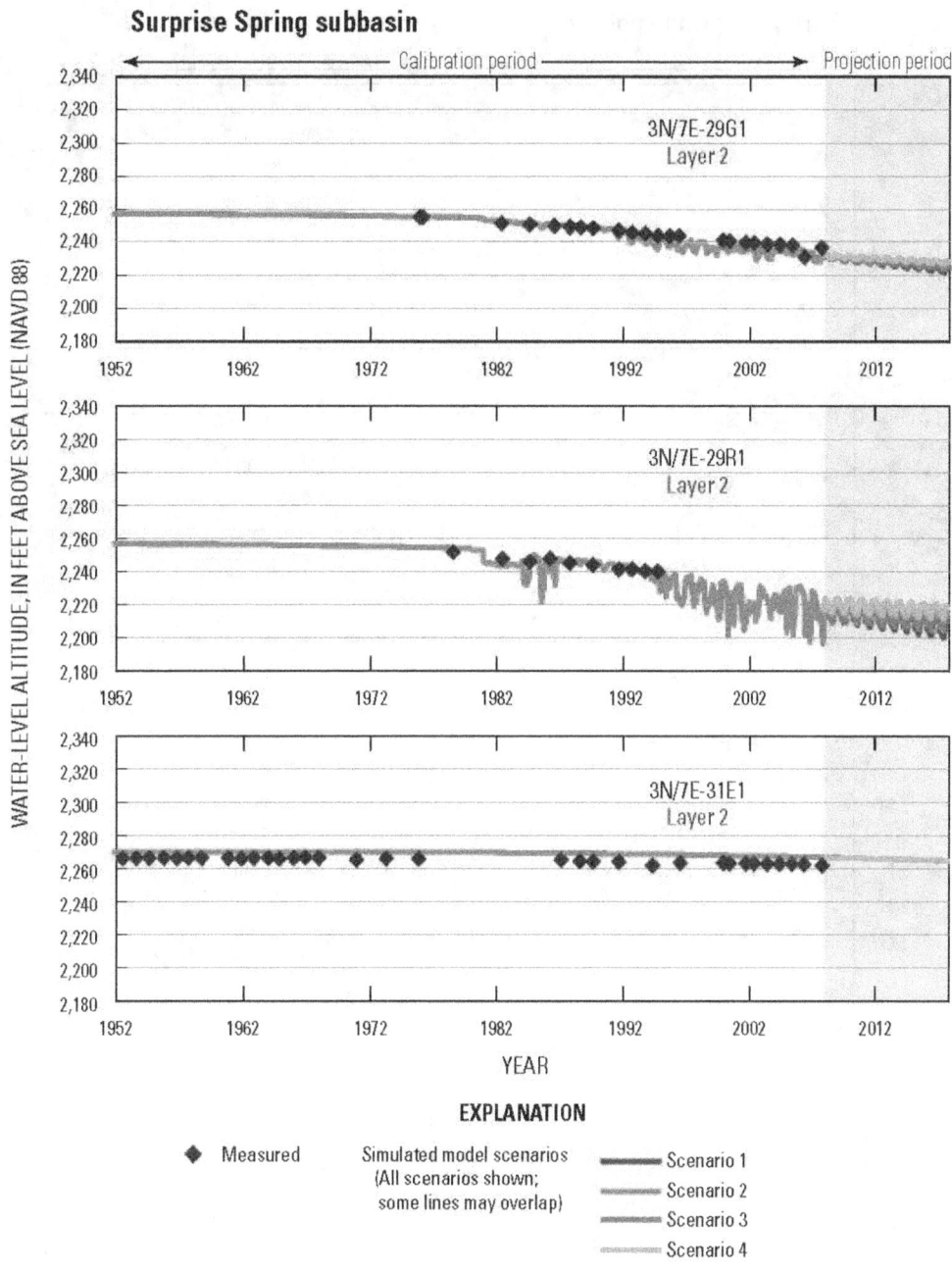

Figure A2   Long-term water-level hydrographs and model simulated hydraulic heads at wells 3N/7E-29G1, 3N/7E-29R1, and 3N/7E-31E1, Twentynine Palms area, California, 1953-2017.

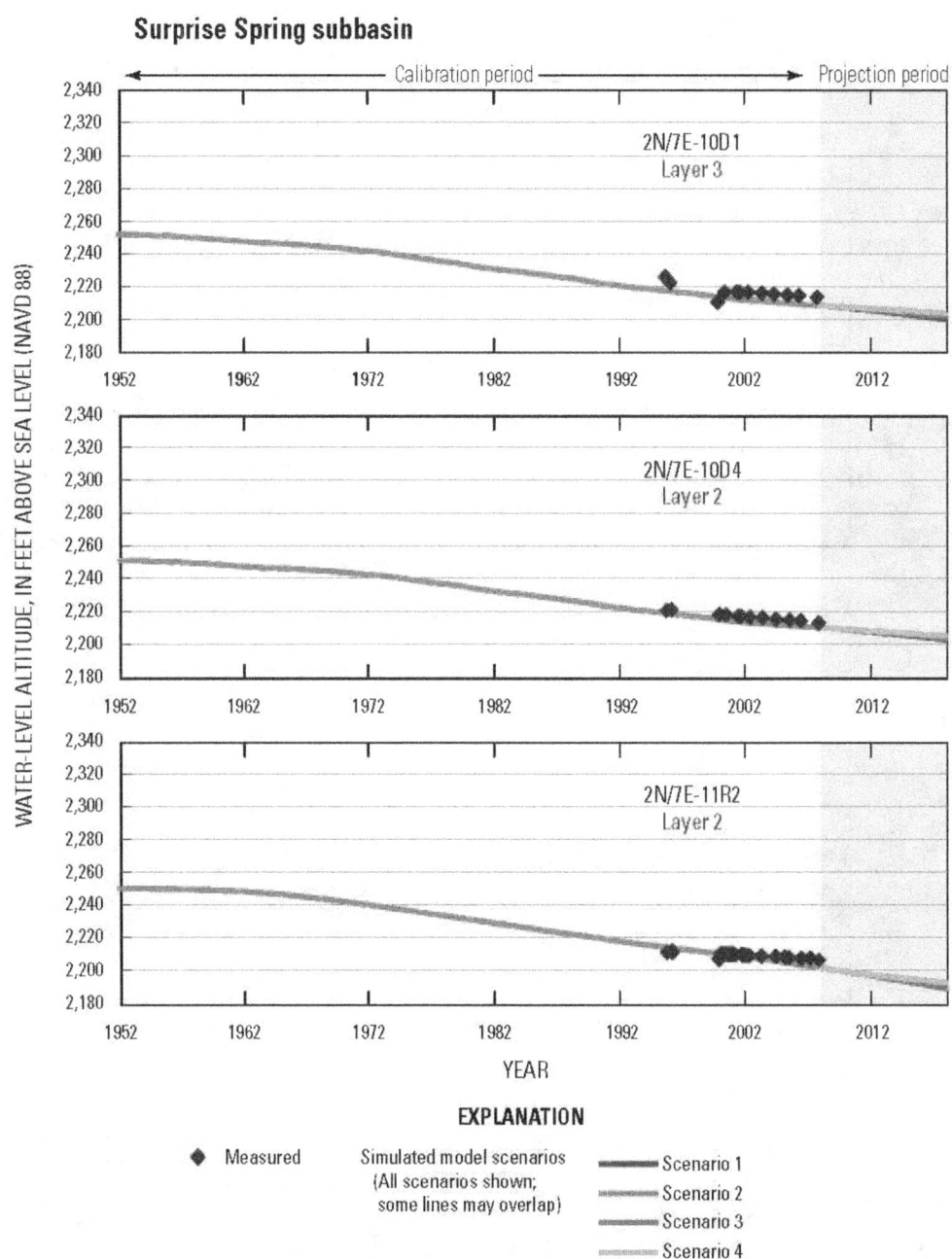

Figure A3   Long-term water-level hydrographs and model simulated hydraulic heads at wells 2N/7E-10D1, 2N/7E-10D4, and 2N/7E-11R2, Twentynine Palms area, California, 1953-2017.

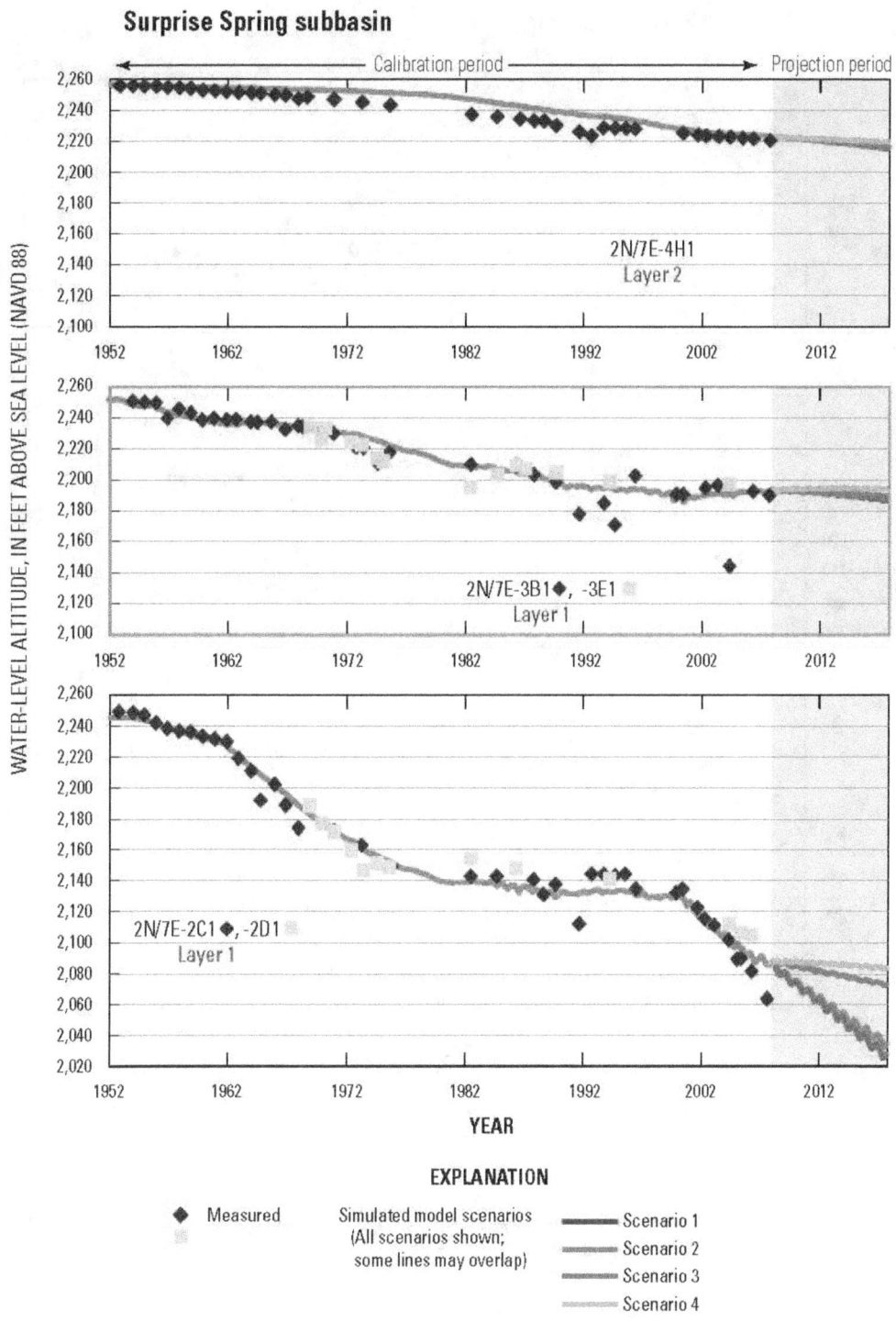

Figure A4   Long-term water-level hydrographs and model simulated hydraulic heads at wells 2N/7E-4H1, 2N/7E-3B1, -3E1, and 2N/7E-2C1, -2D1, Twentynine Palms area, California, 1953-2017.

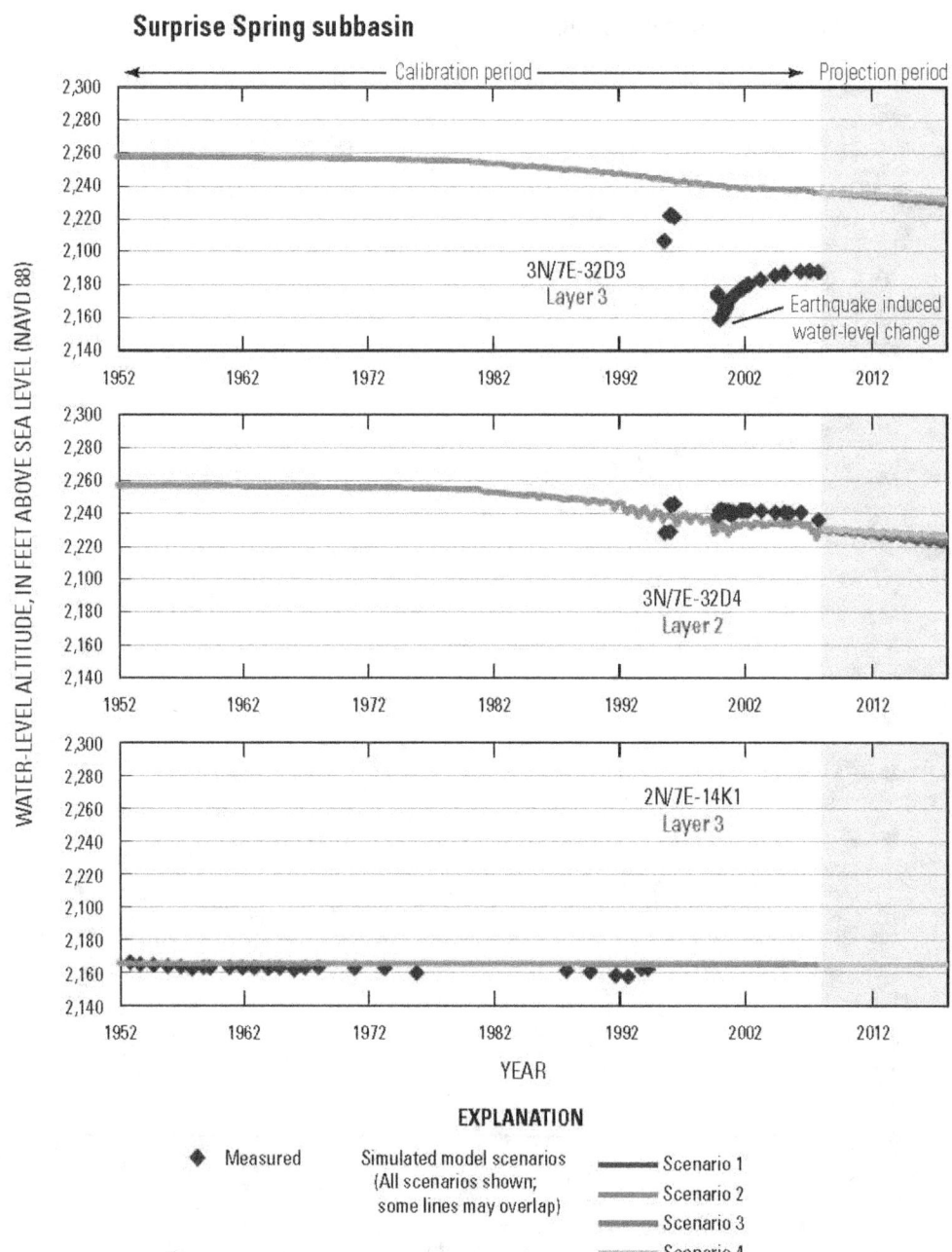

Figure A5   Long-term water-level hydrographs and model simulated hydraulic heads at wells 3N/7E-32D3, 3N/7E-32D4, and 3N/7E-14K1, Twentynine Palms area, California, 1953-2017.

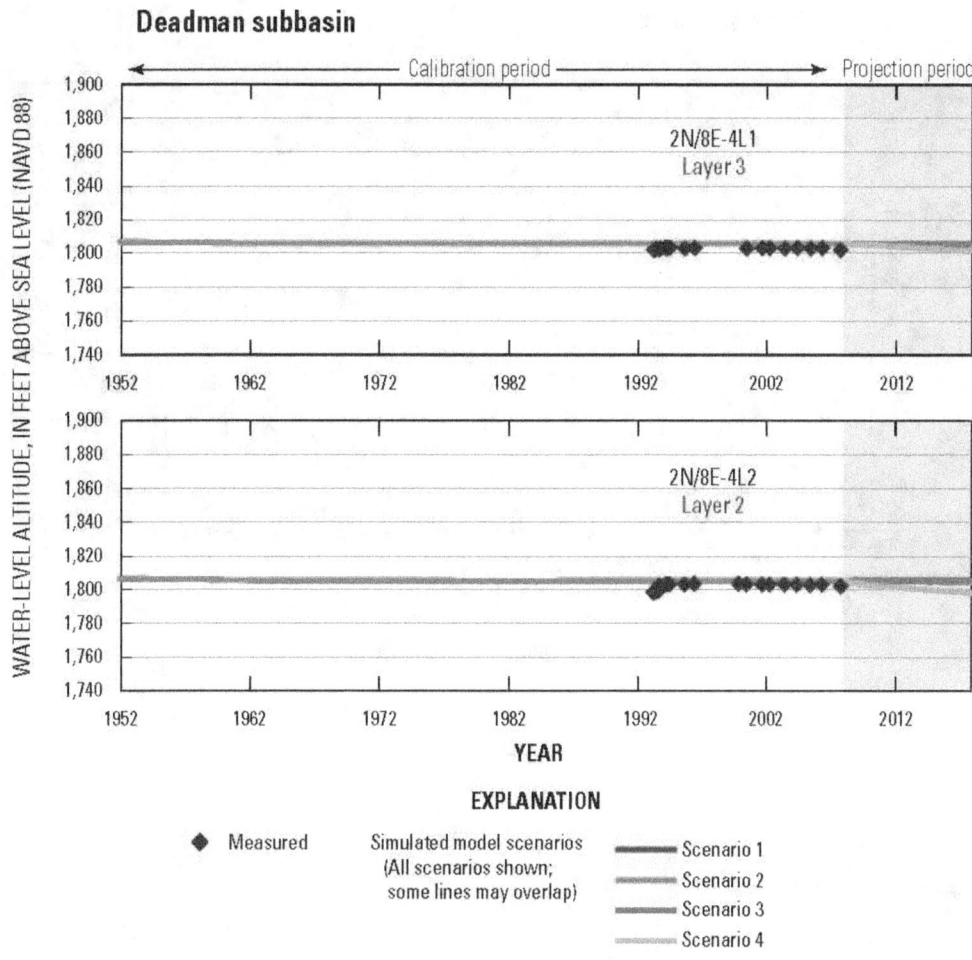

Figure A6   Long-term water-level hydrographs and model simulated hydraulic heads at wells 2N/8E-4L1, and 2N/8E-4L2, Twentynine Palms area, California, 1953-2017.

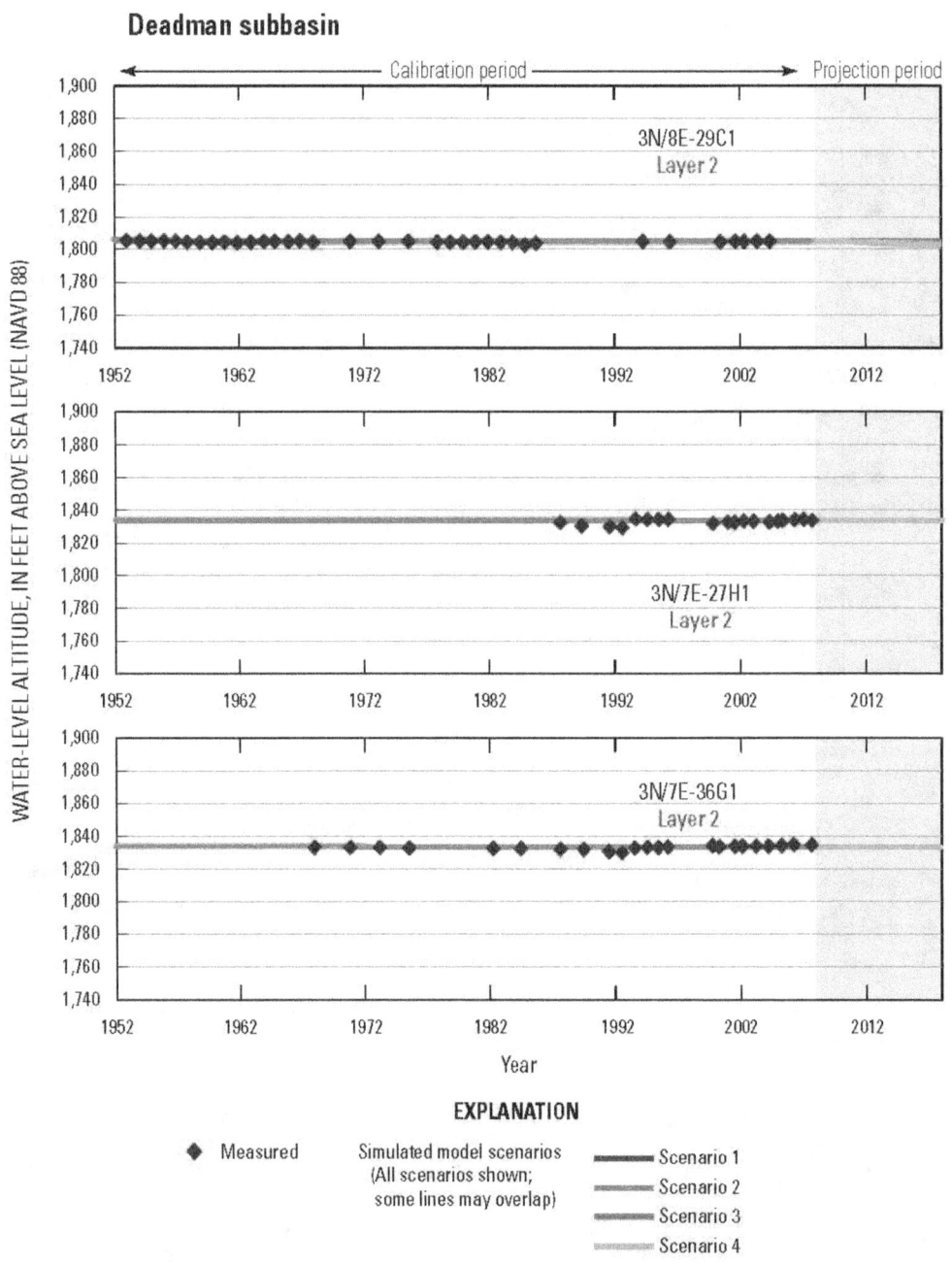

Figure A7   Long-term water-level hydrographs and model simulated hydraulic heads at wells 3N/8E-29C1, 3N/7E-27H1, and 3N/7E-36G1, Twentynine Palms area, California, 1953-2017.

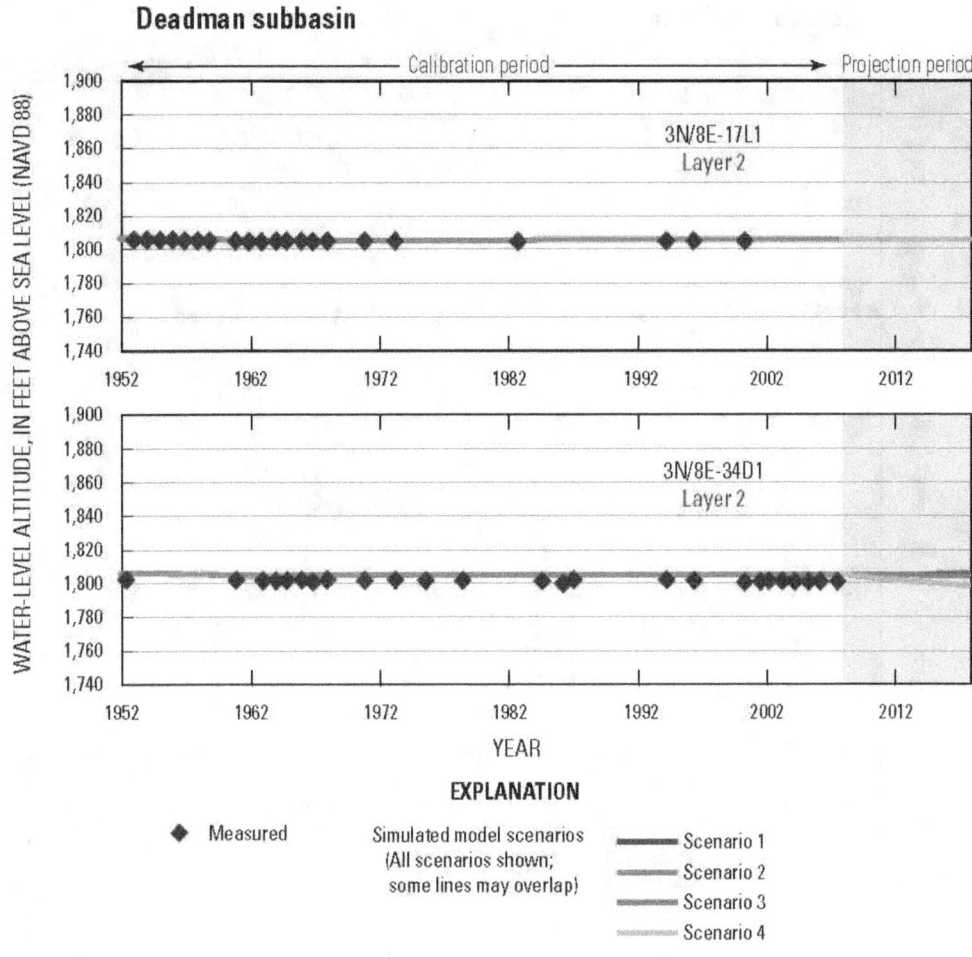

Figure A8   Long-term water-level hydrographs and model simulated hydraulic heads at wells 3N/8E-17L1, and 3N/8E-34D1, Twentynine Palms area, California, 1953-2017.

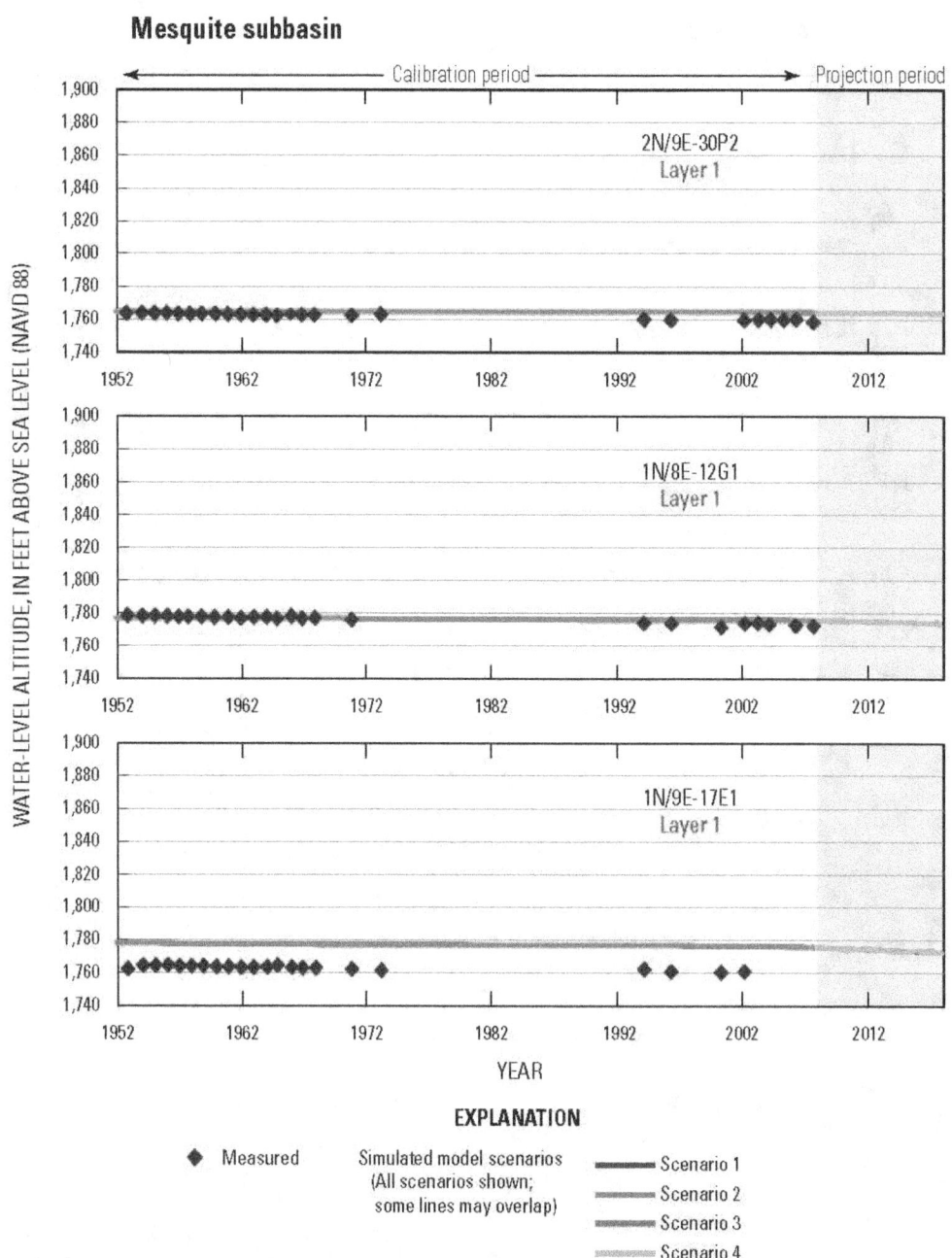

Figure A9   Long-term water-level hydrographs and model simulated hydraulic heads at wells 2N/9E-30P2, 1N/9E-12G1, and 1N/9E-17E1, Twentynine Palms area, California, 1953-2017.

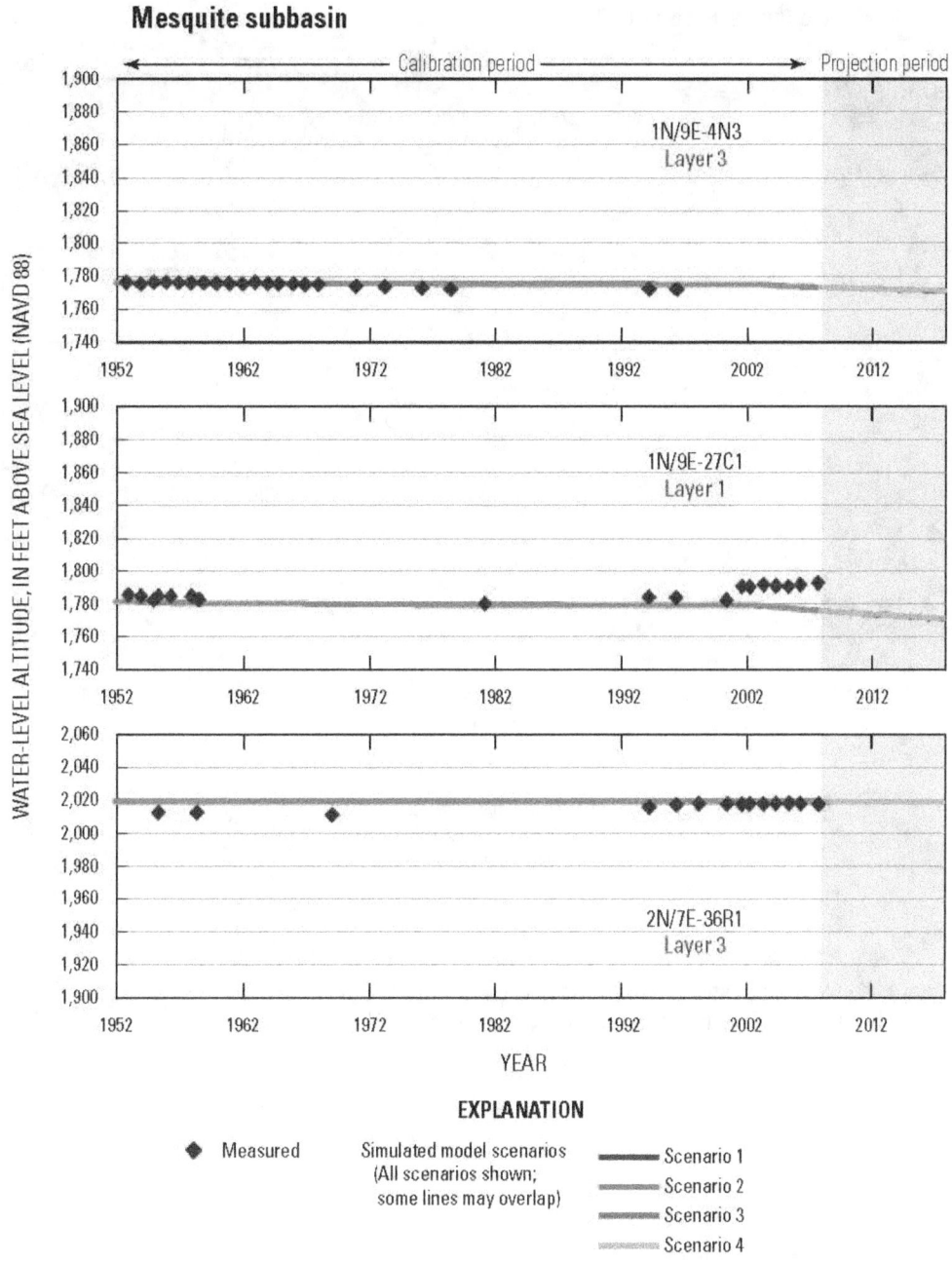

Figure A10   Long-term water-level hydrographs and model simulated hydraulic heads at wells 1N/9E-4N3, 1N/9E-27C1, and 2N/7E-36R1, Twentynine Palms area, California, 1953-2017.

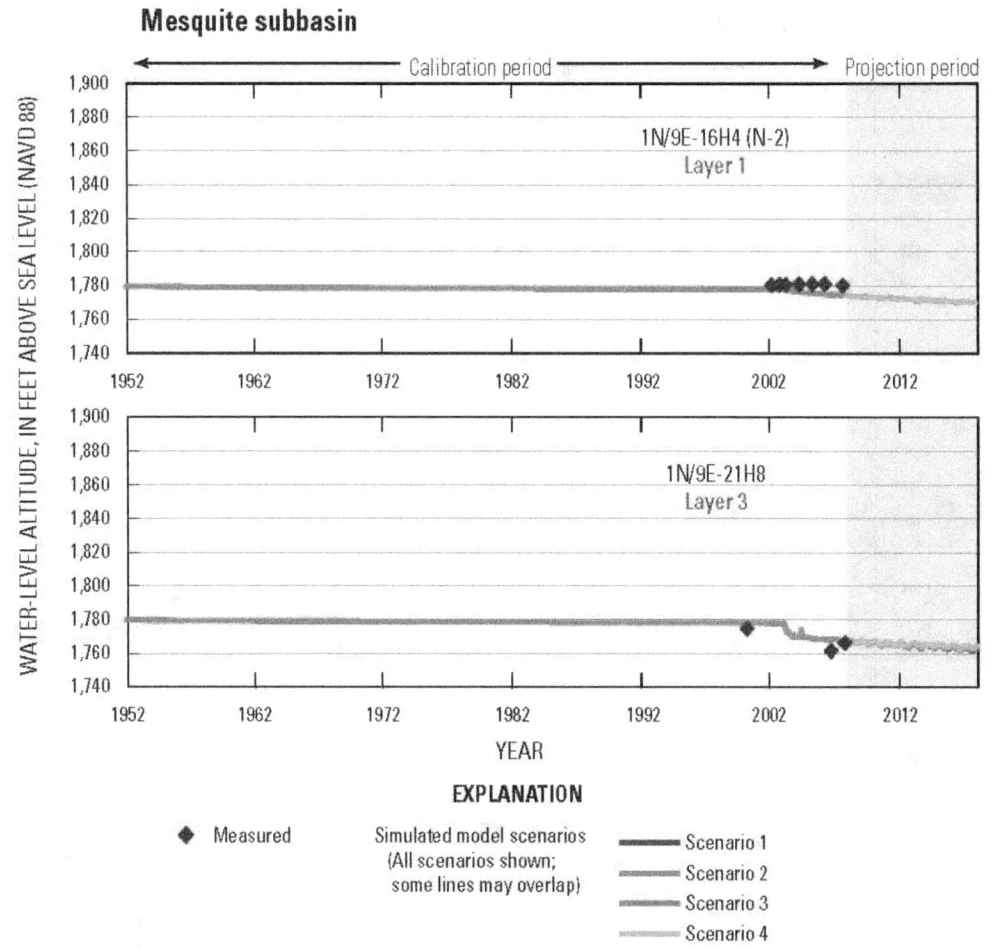

Figure A11 Long-term water-level hydrographs and model simulated hydraulic heads at wells 1N/9E-16H4 (N-2) and 1N/9E-21H8, Twentynine Palms area, California, 1953-2017.

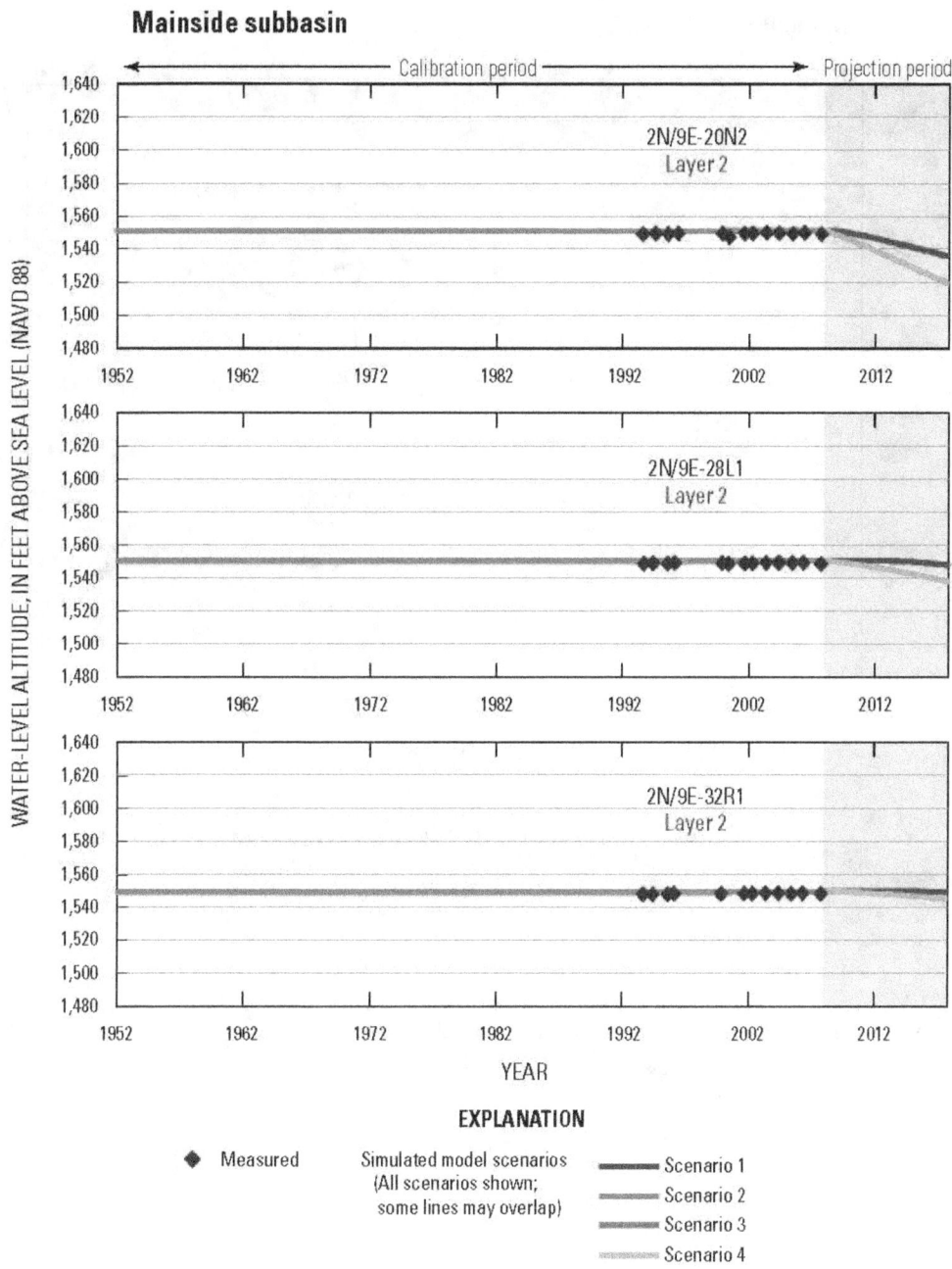

Figure A12   Long-term water-level hydrographs and model simulated hydraulic heads at wells 2N/9E-20N2, 2N/9E-28L1, and 2N/9E-32R1, Twentynine Palms area, California, 1953-2017.

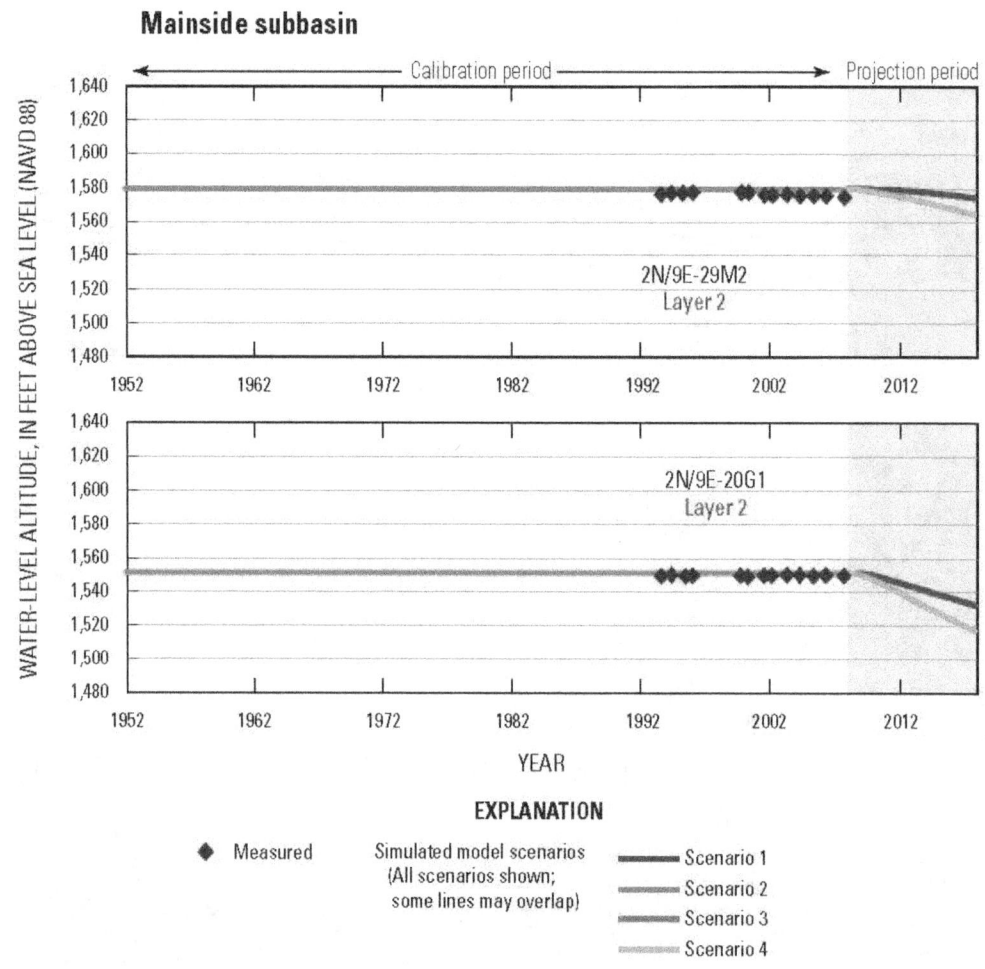

Figure A13   Long-term water-level hydrographs and model simulated hydraulic heads at wells 2N/9E-29M2 and 2N/9E-20G1, Twentynine Palms area, California, 1953-2017.

Appendix B.   Well Construction Information for Selected Wells Used as Observation Points for the Regional Groundwater-Flow Model of the Twentynine Palms Area, California

Appendix B.   Well construction information for selected wells used as observation points for the regional groundwater-flow model of the Twentynine Palms area, California.

[USGS, U.S. Geological Survey; NA, not available]

USGS station ID	State well number	Well depth (feet)	Perforated interval (feet)	Model layer	Aquifer system
Surprise Spring subbasin					
341741116132501	2N/7E–02C1	377	149–377	1	Upper
341732116132701	2N/7E–02D1	532	250–532	1	Upper
341736116141201	2N/7E–03B1	700	260–690	1	Upper
341723116143601	2N/7E–03E1	510	250–510	1	Upper
341720116145601	2N/7E–04H1	420	300–420	2	Middle
341643116144401	2N/7E–10D1	900	880–900	3	Lower
341643116144404	2N/7E–10D4	420	400–420	2	Middle
341601116124802	2N/7E–11R2	485	465–485	2	Middle
341520116130101	2N/7E–14K1	644	450–558	3	Lower
342221116190801	3N/6E–02J1	335	NA–NA	3	Lower
342110116174901	3N/7E–18D1	390	NA–NA	2	Middle
341952116164601	3N/7E–20M1	295	275–295	2	Middle
341912116160801	3N/7E–29G1	348	312–348	2	Middle
341843116155201	3N/7E29R1	600	390–580	2	Middle
341823116175001	3N/7E–31E1	401	NA–NA	2	Middle
341829116164401	3N/7E–32D3	790	770–790	3	Lower
341829116164402	3N/7E–32D4	660	640–660	2	Middle
Deadman subbasin					
341709116090401	2N/8E–04L1	591	571–591	3	Lower
341709116090402	2N/8E–04L2	380	360–380	2	Middle
341909116134101	3N/7E–27H1	587	472–572	2	Middle
341809116115801	3N/7E–36G1	399	384–NA	2	Middle
342037116101101	3N/8E–17L1	455	247–NA	2	Middle
341918116101501	3N/8E–29C1	800	500–684	2	Middle
341823116082201	3N/8E–34D1	396	186–NA	2	Middle
Mainside subbasin					
341449116034201	2N/9E–20G1	440	400–440	2	Middle
341419116040402	2N/9E–20N2	270	250–270	2	Middle
341341116024901	2N/9E–28L1	397	367–387	2	Middle
341340116040501	2N/9E–29M1	410	390–410	2	Middle
341232116032201	2N/9E–32R1	470	450–470	2	Middle
Mesquite subbasin					
341106116090001	1N/8E–09L1	386	346–382	2	Middle
341114116053301	1N/8E–12G1	NA	NA–NA	1	Upper
341141116030901	1N/9E–04N3	495	390–495	3	Lower
341031116041401	1N/9E–17E1	130	NA–NA	1	Upper
340939116021802	1N9E–21H8	500	490–500	3	Lower
340855116013601	1N/9E–27C1	90	NA–NA	1	Upper
341238116114301	2N/7E–36R1	462	305–462	3	Lower
341323116045501	2N/9E–30P2	58	NA–NA	1	Upper